An introduction to total and partial eclipses of the sun and moon

Ian Bruce

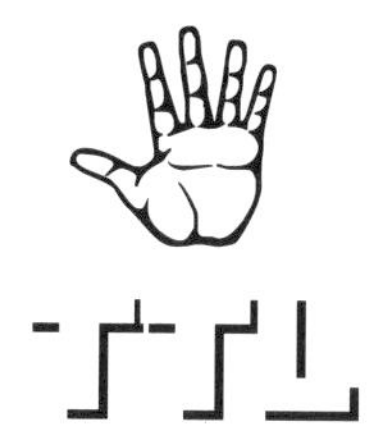

TTL
P.O.Box 200
Harrogate
HG1 2YR

Fax: +44-1423-526035
Email: sales@takethat.co.uk

Acknowledgements: *This book is dedicated to my wife Pauline and my children Hannah and Zachary.*

Also to Colin and Ruth Sherring and Kim and Estelle Vessey.

Thanks to Chris Brown and all at Take That Ltd for their invaluable help and guidance in writing this book.

Finally, a special thank you to Fred Espenak of NASA/Goddard Space Flight Centre for allowing the reproduction of his eclipse data in this book and Dave Dyer for permission to use his photographs.

10 9 8 7 6 5 4 3 2 1

Never look directly at the Sun without adequate protection

Printed and bound in Great Britain.

ISBN 1-873668-63-5

Table of Contents

Introduction

The purpose of this book is to introduce you to the eclipse phenomenon and to help you to view and enjoy them both safely and effectively. To that end, you will learn a great deal about eclipses, including:

- *How an eclipse takes place and what to expect,*
- *What people have believed about eclipses throughout history*
- *The best ways to view the eclipse with or without special equipment*
- *Methods you can use to help children understand and view an eclipse safely*
- *How you can record an eclipse with a camera or camcorder*

In addition, I will introduce you to the subject of amateur astronomy and present a number of appendices which will help you to study the night sky in more detail.

There is no doubt in my mind that an eclipse - whether solar or lunar - is one of the most awesome natural spectacles anyone can ever witness. It is my sincere hope that you will use this book to help you get the most out of them for many years to come.

Enjoy!

Ian Bruce

Chapter One

Eclipse Basics

The phenomenon of the eclipse is one of the most exciting events an individual can ever witness. It doesn't matter if the eclipse is solar or lunar, watching the event take place stirs a deep feeling of awe and wonder which surprises you every time you experience it. Truly, an eclipse is a natural wonder of the universe, and one which makes even the mighty Niagara Falls or the spectacular Grand Canyon pale in comparison.

No matter how unmoved an individual may be by the sight of a meteor or the blazing tail of a comet, an eclipse never fails to grab the rapt attention of a large proportion of the population. Perhaps this is because of the sheer magnitude of the event. After all, we aren't talking here about a mere firework display being enjoyed by just a few hundred people. We're talking about an astronomical event which can be seen, to some extent, by millions of people simultaneously - sometimes in their own back yard.

Despite the fact that humans have been fascinated by eclipses since they first stood upright, most people do not fully understand how and why they occur. The aim of this chapter is, therefore, to provide you with a basic introduction to the language and causes of an eclipse so that the next time you have the opportunity to go and see one, you can truly appreciate the spectacle. It is not intended to be a deep, scientifically-rigorous manual for astronomers and physicists!

Our solar system consists of nine planets which orbit one Sun. One of these planets is obviously our own planet Earth, and the remaining eight are Mercury, Venus, Mars, Jupiter, Saturn, Uranus, Neptune and Pluto. In addition, seven of the nine planets are in turn orbited by one or more moons (natural satellites). The two planets which do not have significant moons are Mercury and Venus.

In simple terms, an eclipse is an astronomical event where one celestial body either totally or partially obscures another. In the case of a lunar eclipse, the Earth passes between the Sun and the Moon. In the case of a solar eclipse, the Moon passes between the Sun and the Earth itself.

Let's illustrate both of these events so that you are perfectly clear about what is happening in each case.

The Lunar Eclipse

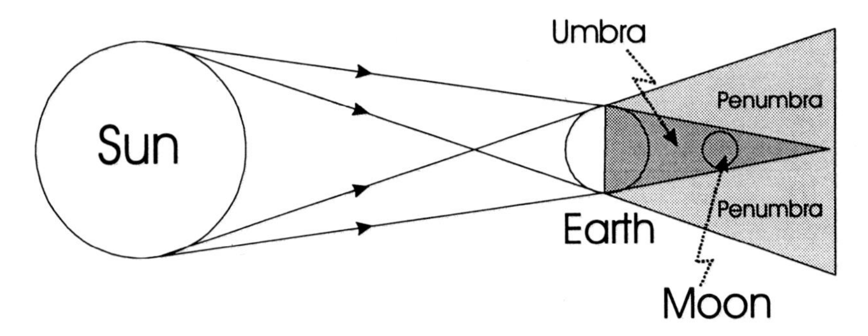

Figure 1
The Moon in the shadow of the Earth

As you can see from Figure 1, a lunar eclipse occurs when the Moon passes into the shadow of the Earth. You will note that the shadow which the Earth casts is actually in two parts (as are all shadows). The darker, inner part is referred to as the *umbra*, and the lighter outer part is referred to as the *penumbra*.

Now obviously, as the Moon passes into the shadow of the Earth, it passes first through the penumbra. This effectively reduces the amount of light which reaches the face of the Moon, thus darkening it from our point of view. As the Moon progresses, it enters the umbra and takes on a dark, reddish appearance. This is because a certain amount of light from the Sun is bent around the Earth (by gravity) and still reaches the Moon. If this were not the case then the Moon would disappear altogether from view whilst in the umbra.

The Moon is the only natural satellite of the Earth. It is 238,000 miles from the Earth and has a diameter of 2,155 miles. Although the landscape of the Moon is often described in terms of valleys, craters and seas, it has no atmosphere or water and is therefore incapable of sustaining known life-forms of any kind. The gravitational pull of the Moon is only one sixth of the gravitational pull of the Earth.

There are three kinds of lunar eclipse:

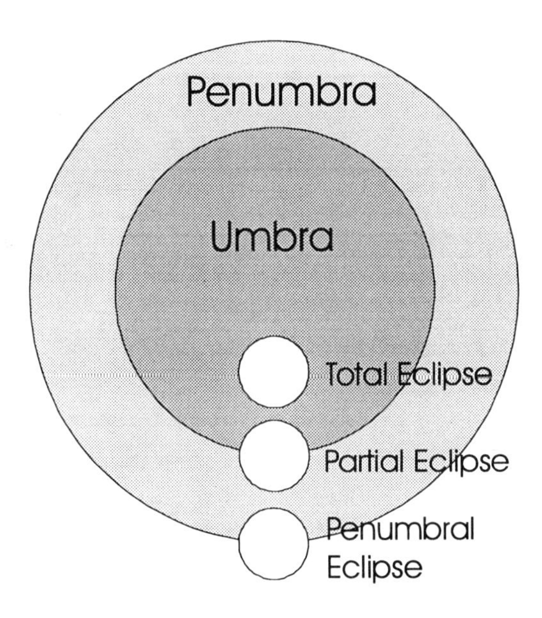

Figure 2
Total, Partial and Penumbral Eclipses of the Moon

- A **Total Eclipse** of the Moon occurs when the entire face of the Moon passes into the umbra of the Earth's shadow. This is the most dramatic kind of lunar eclipse one can witness.

- A **Partial Eclipse** of the Moon occurs when only part of the face of the Moon passes into the umbra of the Earth's shadow.

- A **Penumbral Eclipse** of the Moon occurs when no part of the Moon enters the umbra of the Earth's shadow, but instead passes through the penumbra only.

There are six stages of a total lunar eclipse, and these are more usually referred to as "contacts". Each stage can be defined and illustrated as follows:

First Penumbral Contact (P1)

The stage when the Moon first comes into contact with the penumbra of the Earth's shadow.

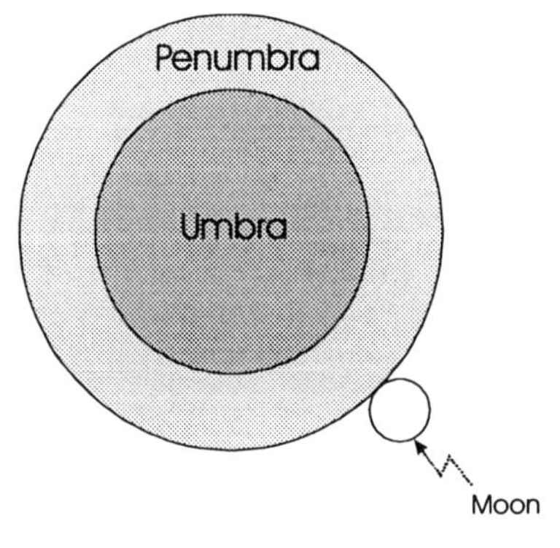

Figure 3-1

First Umbral Contact (U1)

The stage when the Moon first comes into contact with the umbra of the Earth's shadow.

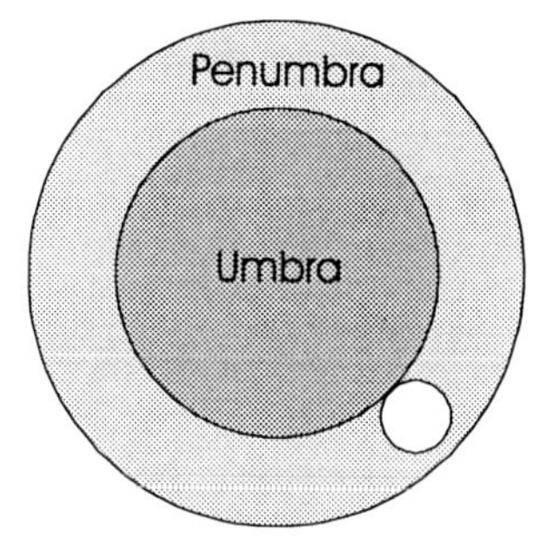

Figure 3-2

Second Umbral Contact (U2)

The stage when the Moon first becomes totally eclipsed within the umbra of the Earth's shadow.

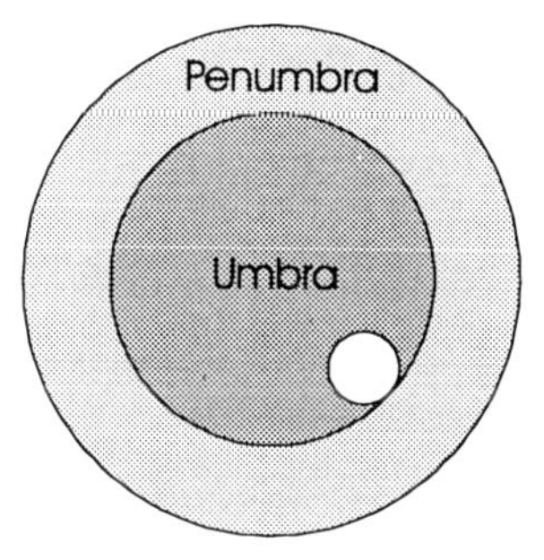

Figure 3-3

Third Umbral Contact (U3)

The stage when the Moon is about to emerge from being totally eclipsed within the umbra of the Earth's shadow.

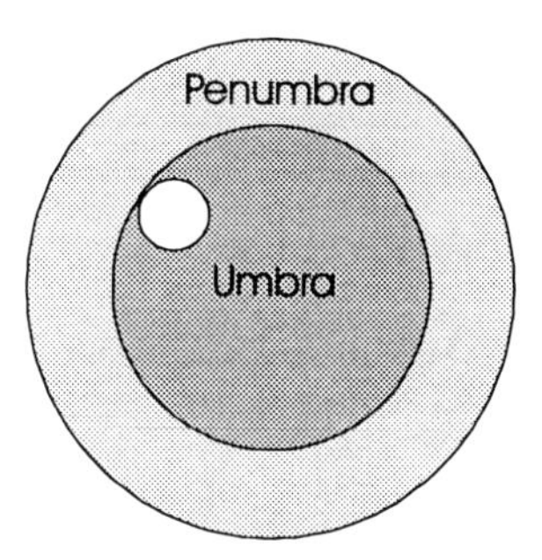

Figure 3-4

Fourth Umbral Contact (U4)

The stage when the Moon has passed out of the umbra of the Earth's shadow.

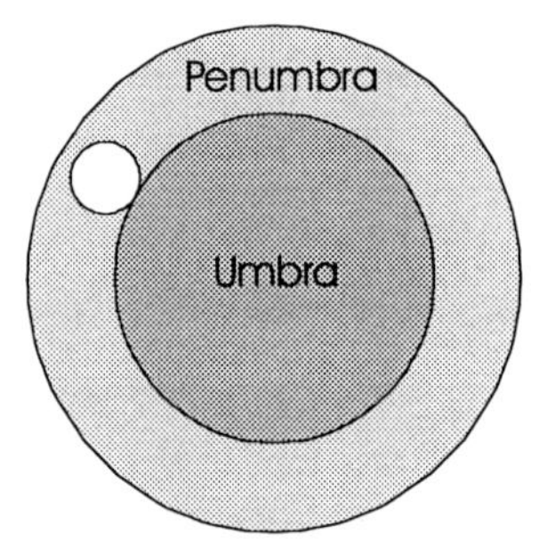

Figure 3-5

Second Penumbral Contact (P2)

The stage when the Moon has passed out of the penumbra.

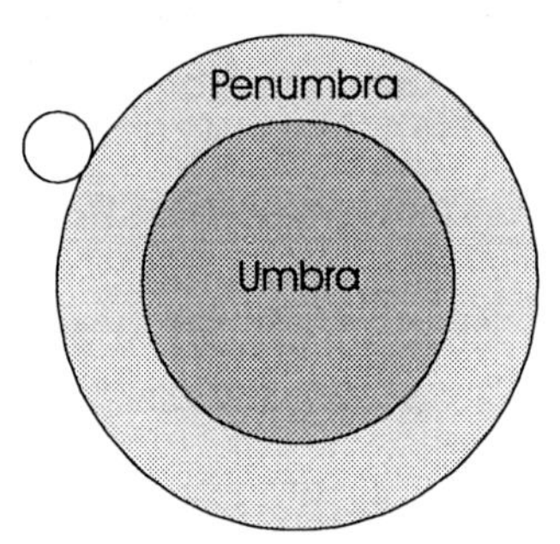

Figure 3-6

It should be noted that a lunar eclipse can only occur when the Moon is full and at a specific point called a node.

Fact Attack

A node is where the orbit of the Moon intersects with the orbital plane of the Earth (usually referred to as the ecliptic). Since this is not a scientific textbook I will not delve into the more technical aspects of inclinations, etc., but suffice it to say that over the course of any one eclipse year (346.6 days) there are never more than seven lunar and solar eclipses combined.

The Solar Eclipse

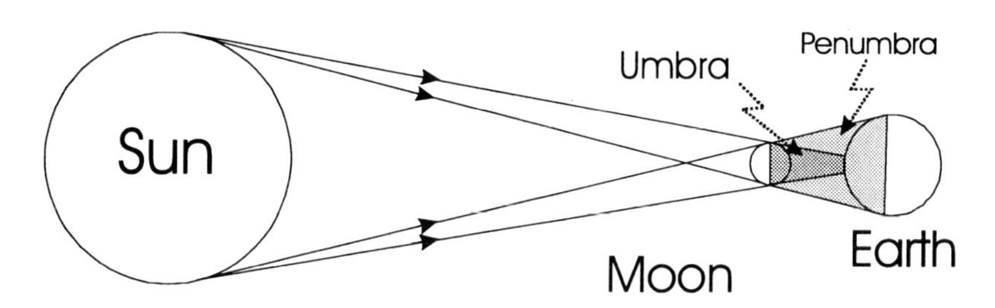

Figure 4
The Moon passes between the Sun and the Earth.

A solar eclipse can only be seen during the New Moon phase. This illustration shows clearly that when such an eclipse takes place it does so because the Moon passes between the Sun and the Earth and the shadow of the Moon appears on the Earth's surface. From an Earth-bound perspective the "kind" of eclipse witnessed depends on the position of the viewer in relation to the penumbra and umbra of the shadow.

- A **Total Eclipse** of the Sun will be seen by anyone who is situated in the umbra part of the shadow.
- A **Partial Eclipse** of the Sun will be seen by anyone who is situated within the penumbra part of the shadow.
- On some occasions only the penumbral part of the shadow cast by the Moon appears on the Earth's surface. This obviously means that no total eclipse of the Sun can be witnessed, but those within the penumbra may still view a partial solar eclipse.

Because the Moon orbits the Earth in an elliptical manner, the perceived size (of the Moon when viewed from Earth) can vary. Such a variance in perspective results in two different variations of solar eclipse:

- An **Annular Eclipse** of the Sun occurs when the Moon appears to be smaller than the Sun. Because the Moon cannot block out the whole of the Sun in such an event, a glowing ring (or annulus) of bright light remains visible in the sky. This makes viewing some of the more spectacular total-eclipse phenomenon, such as the corona (see later in this chapter), very difficult or even impossible. Having said that, viewing an annular eclipse is better than viewing none at all, so if you get the opportunity to do so, take it.

The Sun is the star at the centre of our solar system, around 93 million miles from the Earth, with a diameter of 865,000 miles and a surface temperature of approximately 5,500° Celsius. The energy of the Sun (and any other star for that matter) is created through a process of nuclear fusion at its core which turns hydrogen into helium. Scientists have estimated that around half the hydrogen at the core of our Sun has already been converted into helium over its life of 4.7 billion years. This means that the Sun is nearly half way through its estimated lifespan of around 10 billion years.

- An **Annular-Total Eclipse** of the Sun occurs when the Moon appears to be smaller than the Sun (as in an annular eclipse), but then appears larger as it progresses, creating a very short total solar eclipse. This type of eclipse is quite rare.

Just as there are several key stages in a lunar eclipse, so there are also key stages in a solar eclipse. Again, these are referred to as "contacts", and there are four of them, as follows (note that the illustrations show an annular eclipse so that the sun's position relative to the Moon at each contact stage can be seen):

First Contact (P1)

This stage is when the Moon first begins to obscure the disk of the Sun.

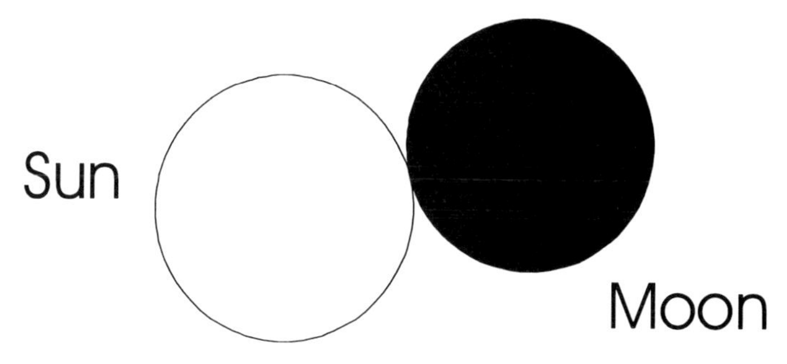

Figure 5-1

Second Contact (U1)

This stage is when the Sun enters the phase of totality.

Figure 5-2

Maximum Eclipse

This is not a stage in itself, but is merely the midpoint between U1 and U2.

Figure 5-3

Third Contact (U2)

This stage is when the Sun is about to leave the phase of totality.

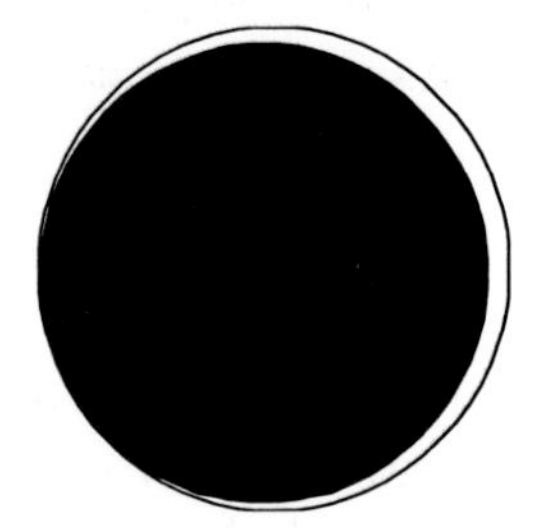

Figure 5-4

Fourth Contact (P2)

This stage is when the Sun is no longer obscured by the Moon.

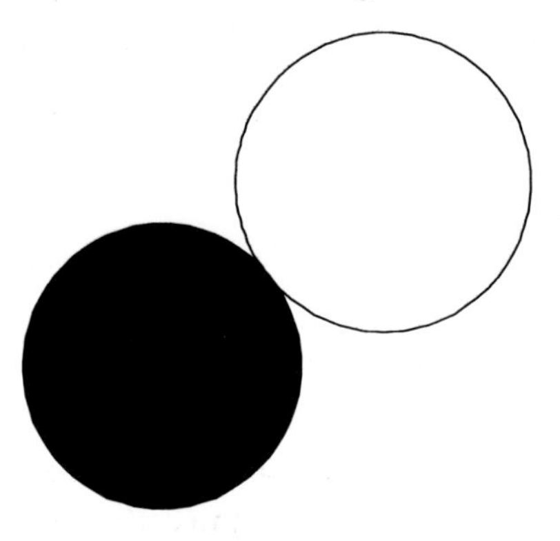

Figure 5-5

Special Solar Eclipse Events

You now know the key stages of a solar eclipse, but there are also some special solar eclipse events which are well worth looking out for.

Shadow Bands

The first thing you should look for is not in the sky at all - but on the ground. As the eclipse approaches second contact, look to the ground for something called *Shadow Bands*. These are caused by the sliver of sunlight visible near second contact being bent by the Earth's atmosphere, and they appear as ripples of light which appear to be moving across the landscape - often quite quickly.

Tree Images

If you are observing the eclipse from a site which has a number of trees, observe these from time to time throughout the event. Trees often act as "nature's pin-hole projectors" in that light from the Sun is directed through tiny gaps between some leaves and small projected images of the eclipse appear on others. Whilst such an occurrence may not sound particularly spectacular, a tree which naturally projects several images of the eclipse at once is an intriguing sight which you will be glad to have noticed.

Baily's Beads

Next, look for *Baily's Beads*. These appear during second contact when the Moon obscures the Sun. Light from the sun's *photosphere* seeps through the mountains and valleys of the lunar landscape and the result is something which resembles a "string of bright beads." This description is not my own, I hasten to add. It belongs to one Francis Baily, the man who is credited with bringing the phenomenon to the attention of astronomers after witnessing it during an 1836 annular eclipse.

Diamond Ring

As the Baily's Beads begin to disappear, there is often a point where just one shaft of photospheric light is visible. This appears as though it were some kind of flare, and because the photosphere of the Sun looks something like a ring at this point, gives the whole eclipse the look of a *Diamond Ring*. For many, this feature is the most stunning of the whole eclipse. What do you think? Page 16 shows an example from California in 1991.

The Diamond Ring seen in Hawaii and California on 11th July 1991.
(Photographed by Dave Dyer, 1/250 second, f5.6, asa 100, 400 mm lens)

The Corona seen on 11th July, 1991. Totality lasted an amazing six minutes. (Photographed by Dave Dyer, 1/15 second, f5.6, asa 100, 400 mm lens)

Corona

Following the appearance of the Diamond Ring, the *corona* takes centre stage and blooms out from behind the Moon. If you look closely as the eclipse enters the phase of totality, you may be able to see the *chromosphere* of the Sun. This is a layer of gas which exists between the photosphere (visible during the Diamond Ring stage) and the corona, but it is usually only visible for a few precious moments before it gives way to the full bloom of the corona itself. It is impossible to even describe something as magnificent as the corona, so I won't even attempt to do so. All you need to know is that it appears right between second and third contact. An example from California in 1991 is shown on the previous page (p17). The moon's shadow was 200 miles wide on this occasion. When combined with a favourable relative spin of the Earth and Moon, it lead to a totality which lasted for more than six minutes.

Prominences

Another thing you should look out for are what are known as *prominences*. These are gigantic fountains of gas which reach high into the corona and take on a glowing reddish appearance. Prominences can be visible throughout an eclipse so make sure you pay attention.

As the eclipse reaches third contact, these phenomena will be repeated, though obviously in reverse order. The corona will fade and the chromosphere may be visible once again. Then watch for the appearance of the Diamond Ring, followed by Baily's Beads.

In a later chapter we will discuss how you can record a lunar or solar eclipse with a sketchpad, camera or camcorder. But in the meantime, let use move on from our explanation of how an eclipse takes place and begin to explore a subject which is as relevant today as it ever was - the mythology of eclipses...

Chapter Two

The Mythology of Eclipses

In the last chapter we saw how an eclipse occurs. The event is easily understood in the modern age because we know where our planet is in relation to the Sun, the Moon and other planets in our solar system. In other words we can accept the phenomenon for what it is - the simple obscuring of one celestial body by another.

Previous generations, however, were not always so fortunate. The modern scientific explanation for why an eclipse occurs was not available for much of the last four thousand years, and so it is not surprising that people turned to home-spun myths, legends and beliefs to try and account for what was making the Sun vanish or the Moon turn crimson red.

One of the most famous stories concerning ancient eclipses and the kind of mythology which went along with them is the tale of Hsi and Ho. The backdrop to this story is that the common belief in China around 2,500 BC was that an eclipsed Sun was actually being consumed by some gigantic dragon. The emperor at the time had appointed two astrologers, Hsi and Ho, to predict such consummations and carry out certain quasi-religious rites to ensure that the Sun survived the attack.

One on particular occasion, a solar eclipse took place and neither of the astrologers had predicted the event. Enraged at their lack of responsibility, the emperor went in search for Hsi and Ho and found that they were drunk on wine. He ordered their execution and both were beheaded - despite the fact that the Sun survived the dragon's attack without their help.

The belief that a dragon is responsible for an eclipse was widespread in the ancient East, and even today some people bang drums and dance in the street in order to try and chase the beast away. Of course, in modern times this is more of a tradition than a deeply held belief, but it does illustrate how mythology can be passed on from generation to generation despite scientific evidence to the contrary (i.e. the Sun always returning).

Sun and Moon Worship

Although Sun worship today means little more than splashing on a bottle of UV protection cream and spending a couple of hours lying on a beach in order to get a better tan, for ancient peoples the phrase described something much more serious. Sun worship was a very real religion for them.

The idea that otherwise reasonable men and women could actually worship the Sun seems a little preposterous today, but in ancient times people believed that the star at the centre of our solar system was actually a god. Their reasons for drawing such an astonishing conclusion are actually fairly straightforward. The Sun is the biggest thing in the sky, and this suggests that it has great power. The Sun helped the ancient peoples to grow their crops. It provided them with heat and light. It could be relied upon to greet them each morning. It accompanied them whenever they travelled - no matter how much distance they covered. Thus the Sun was a powerful being - the Great Provider and Protector - and was more than worthy of being worshipped.

This widespread belief in the Sun as a god was so strong in ancient times that temples were built, liturgies developed and rituals performed so that the people could pay this god its due respects. In addition, the Sun-god was often personified by being given a name. The ancient Greeks called their god *Helios*, whilst the ancient Romans opted for *Sol*.

Moon worship was never as marked as Sun-worship. However, worship of the Moon did occur and the belief that the Moon was a god in its own right was quite widespread in the ancient world. Phases of the Moon were even linked with states of mental health (or otherwise) in the population, hence the modern words lunacy and lunatic have their roots in the Latin word *Luna* (Moon).

With so many people worshipping the Sun and Moon as gods, it is no wonder that eclipses were widely viewed in the ancient world as something of a bad omen. It seemed to represent the "drawing back" of the god from its followers - a gesture of displeasure with the people who worshipped it. Because of this belief, a plethora of rites, ceremonies and so on were carried out whenever an eclipse took place.

Obviously, these actions appeared to work because no eclipse is ever permanent, and so Sun worship (and to a certain extent Moon worship) thrived and prospered until genuine scientific understanding of the event evolved and exposed the beliefs for what they truly were - mythical.

The Eclipse as an Omen

Despite the decline of Sun and Moon worship in a religious sense, the idea of a solar or lunar eclipse being an omen is one which has proved very difficult to shake off. Even in the modern world there are many people who believe that eclipses, comets and other such celestial spectacles are "signs" from higher powers, "warnings" that something catastrophic is about to take place or - in the current era at least - "calls to enlightenment" from advanced alien civilisations.

Some people believe that eclipses are bad omens. Many think that there is a link between eclipses and Earthquakes, wars, famines and even stock market declines. Even as late as 1948, a Korean political election was postponed because a total eclipse was scheduled to take place on the same day. Did the organisers of this election unreservedly believe that the eclipse would actually affect the result? Probably not, but they still believed it enough to "play on the safe side" and re-schedule the event.

The problem with the belief that an eclipse is a bad omen is that the belief itself can bring about disaster. Apparently, in 840 AD, Emperor Louis was so taken aback by a total eclipse of the Sun that he died of fright. But it wasn't the eclipse that killed him - it was his own fear which came from a basic misunderstanding of the phenomenon.

Others, however, have turned the mythology of eclipses to their own advantage. For example:

When stranded on the island of Jamaica with his crew, Christopher Columbus consulted an astronomical almanac and discovered that a lunar eclipse was scheduled to take place in a matter of days. He decided to use this information for his own purposes by taking advantage of the native's superstitions. He advised the Jamaicans that unless they provided food for himself and the crew, his God would make the Moon turn dark red. When the time arrived, the lunar eclipse took place exactly as Columbus had warned. The natives were so terrified they immediately presented the visitors with food - and continued to keep them well looked after until Columbus and his crew were rescued.

There are still others who believe that an eclipse is a good omen - that it heralds a new transition in the evolution of the human race, a breakthrough in technology or a new dawn of spiritual enlightenment.

The degree to which people attribute the power of an eclipse is normally directly related to the type of eclipse in question. Lunar eclipses, which are less spectacular than their solar counterparts, are seldom believed to be particularly strong omens. A total solar eclipse, however, arouses a deeper sense of expectancy. It is as though the event itself is not enough, and that there has to be some "meaning" for something so spectacular to take place. Thus, as long as people continue to think in this manner, the belief in an eclipse as an omen (whether good or bad) will continue to be propagated.

Eclipses and the Bible

The Bible is well known for its apocalyptic prophecies, and even a casual reader of this ancient compilation of writings would find it easy to present extracts which appear to link the eclipse phenomenon with disaster and tribulation. Consider the following selection:

"*And I looked when He broke the sixth seal, and there was a great Earthquake; and the Sun became black as sackcloth of hair, and the whole Moon became like blood...*"
Revelation 6:12

"*And there will be signs in Sun and Moon and stars, and upon the Earth dismay among nations, in perplexity at the roaring of*

the sea and the waves, men fainting from fear and the expectation of the things which are coming upon the world; for the powers of the heavens will be shaken."
Luke 21:25-26

"But immediately after the tribulation of those days the Sun will be darkened, and the Moon will not give its light, and the stars will fall from the sky, and the powers of the heavens will be shaken."
Matthew 24:29

"And I will display wonders in the sky and on the Earth, Blood, fire and columns of smoke. The Sun will be turned into darkness, And the Moon into blood. Before the great and awesome day of the Lord comes."
Joel 2:31

In addition consider the account of the crucifixion of Jesus of Nazareth:

"And when they came to the place called The Skull, there they crucified him and the criminals, one on the right and the other on the left... And it was now about the sixth hour, and darkness fell over the land until the ninth hour, the Sun being obscured; and the veil of the temple was torn in two."
Luke 23:33,44-45

These extracts may all seem to refer to something similar to the eclipse phenomenon, but that is all that can be said accurately about them. Whether the remaining text of the verses above is accurate or not and if this is a good or bad thing depends largely on your point of view.

- Fundamentalist believers in the Bible would see eclipse phenomenon largely as a good omen for themselves, for the extracts indicate that some kind of eclipse will herald the return of their Messiah. They would also, however, point out that it is a bad omen for non-believers, who will be punished for their non belief during some kind of tribulation period.

- Liberal believers in the Bible would see the eclipse phenomenon as being used in a more poetic context, and would seldom take such prophecies literally. Instead, they would view such ideas as the tribulation, the returning Messiah and the end of the world as being pure metaphor which are open to deeper and more spiritual forms of interpretation.

- Sceptics who believe that the Bible is nothing other than an occasionally insightful and inspirational collection of ancient writings, would take the whole idea of prophecy and eclipses with a generous pinch of salt. They would be unlikely to believe in omens on any level, let alone based on a text which is primarily religious in nature.

The above conclusions make it clear that as long as there are people with different points of view, there will always be a few who see an eclipse as being the fulfilment of a Biblical prophecy. Even if an eclipse passes uneventfully as far as the apocalypse is concerned (and all have done so far), these people will still believe that we are one eclipse closer to heaven or hell, depending on which side of the fence you sit on with regards to your view of the Bible and its contents.

Astrology

No chapter on the mythology of eclipses would be complete without at least a mention of astrology. As you probably know, astrology is a belief system which is based on the notion that our daily lives here on Earth are influenced - to a greater or lesser degree - by the movements of celestial bodies. The logic of this particular belief system seems to be that if the Moon governs the tides then all celestial bodies must have some influence over everything elsc on planct Earth. This influence, proponents say, is not limited to the lives of human beings, but also affects financial markets, business success and even world peace.

Like all types of eclipse mythology, it is futile to argue against astrology on the basis of science. Those who believe in it will continue to believe in it no matter what the established scientific community say, so it looks like astrology will be around for many years to come. Since this is likely to be the case, here is what you might expect astrologers of various persuasions to say about an eclipse...

- *A personal astrologer who draws up charts for an individual based on the time, date and place of his birth, would say that an eclipse heralds a time of unexpected change. Perhaps there will be a birth, marriage or promotion in his life. An eclipse may also be said to bring a great amount of self-realisation or clarity into his life. The event signifies the end of one way of life and the beginning of a new one. A "rebirth" if you will.*

- *A financial astrologer who draws up charts for stocks, commodities, financial indices, and so on, might interpret the eclipse as being a sign of market reversal. If the financial markets are currently doing well then one might expect the bull market to end and a bear market to take over. If the financial markets are suffering, then the eclipse might lead you to expect a strong bull market to establish itself. Again, the bottom line is that the eclipse signals an "all change" and thus indicates that boom will turn to bust, or vice versa.*

- *A global astrologer who draws up charts for entire groups or even populations would look at the eclipse and predict change on a massive scale. Maybe there will be a change in leadership or government policy. Maybe arguments and wars could start where none existed before. Or maybe*

> *existing arguments and wars could come to an end. Whatever happens, you can bet your last penny that it's all part of some "greater evolutionary process" and that we shouldn't worry too much.*

As we have seen, the mythology of eclipses as omens, fulfilment of Biblical prophecies or astrological "all change" warnings are as common today as the old "Sun-eating-dragon" tales were in the ancient East. Add to this the fact that many millions of people openly admit to a belief in UFO's, alien intelligence, higher powers and other such phenomenon and it looks certain that some kind of eclipse mythology will be with us for many years to come. The myths themselves may change in content, but they will never totally die.

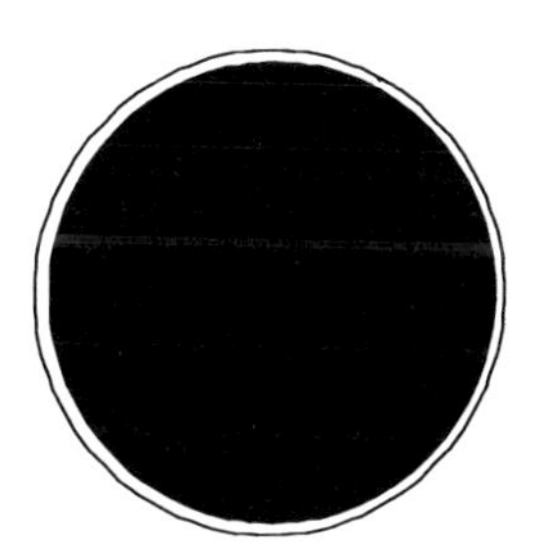

Chapter Three

Preparing to View the Eclipse

The amount of preparation you will need to do to view an eclipse effectively depends on three things:

- *The type of eclipse you intend to view.*
- *The location of the eclipse in relation to your home base.*
- *The quality of viewing experience you want to have.*

Generally speaking, the rarer the type of eclipse, the more travelling you intend to do and the better quality of experience you want, the more preparations you will have to make. Conversely, if all you want is to take a look at a lunar eclipse in your own back yard, the preparations you will have to make are negligible.

In this chapter we will discuss all you need to know about preparing to view an eclipse, from how to choose the best place to view an eclipse to making the necessary travel arrangements and getting ready for departure. To make this information even more useful, a thorough checklist has been provided at the end of the chapter to help ensure that everything goes as smoothly as possible.

Choosing a Viewing Site

The first step in preparing to view an eclipse is to find out which sites are most likely to provide the best experience. In the case of a lunar eclipse it may be that you will be able to get a good view in your own neighbourhood. In the case of a solar eclipse, however, the best site will undoubtedly lie along the path of totality - and this could be many hundreds of miles away.

Wherever you decide to view from, you should ensure as far as possible that the site itself will have as little light pollution as possible. This is particularly important if you intend to photograph or film the event, as you will see in Chapter Six. Even though a solar eclipse takes place in the day, the sky obviously gets dark during the eclipse itself and if the locality has a lot of automatic street lamps or other man-made forms of illumination, these can detract from both the experience and the quality of any photographs or video footage you may take. An open site such as a field with minimum light pollution would therefore be ideal.

Of course, if a particular eclipse is likely to attract an enormous amount of attention then it can be very difficult, if not impossible, to locate such a site. In these situations one has to settle for the best viewing site on offer. Note, however, that if you are willing to pay, there are alternatives to viewing from land.

One alternative which is becoming increasingly popular is to view a solar eclipse from a cruise ship. There are many companies all over the world which organise special short trips specifically for the purpose of allowing people to experience the event away from land. Obviously there is a lot less light pollution and general crowding at sea, so it is not surprising that many serious eclipse photographers and home video enthusiasts prefer the cruise ship experience to any other.

Although viewing from a cruise ship does have advantages over viewing on land, it still does not take away the very real possibility that the event can be obscured by cloud cover. For this reason, people who can afford it choose another alternative: viewing from the air. They do this by booking places on specially chartered "eclipse flights" which an increasing number of air travel companies are offering. This ensures that the passengers can get a clear view of the eclipse because they can rise above any cloud cover. Unfortunately, this option is not so good for photographers and so on because taking clear pictures or video footage through the windows of a plane can be extremely difficult. The bottom line is, therefore, that eclipse flights tend to be favoured by people who are happy to simply witness the event without necessarily recording it to any professional extent.

Despite both of these alternatives, the vast majority of eclipse watchers tend to do so from land. Not only is it less expensive than the alternatives of viewing from the comfort of a cruise ship or an aeroplane, it is also far easier to organise.

One of the major factors you must take into consideration when choosing where to view from is the weather. The fact is that some countries have far more cloud cover on an average day than others. It makes sense, therefore, that if you can afford to do so you should give preference to viewing from an area where the weather is likely to be good and the skies clear. Of course, the weather in any place can never be guaranteed, but playing the probabilities may mean the difference between experiencing the glory of a solar eclipse and experiencing nothing more than getting drenched in a sudden downpour.

As we have said, in the case of a solar eclipse, the path of totality (weather permitting) offers the best view. You should note, however, that it is not vital that you view from the centre line of the path of totality. The path of totality itself could easily be fifty or one hundred kilometres wide, and anywhere within this path will be likely to provide a spectacular view. Knowing this is important, particularly if official viewing sites and places of accommodation along the centre line of the path of totality are already fully booked.

Booking Your Place

Once you have decided on where you want to view the eclipse from, if you are aware that you will need to travel to this site, you need to start making travel and accommodation arrangements as soon as possible. This is because if you deem the eclipse worthy of travel, so will many thousands of other people. Local hotels and guest houses will therefore get fully booked fairly quickly, so it is best to reserve your place as soon as you have made the decision to travel to the area. If you intend to use public transport then the same applies for booking your ticket.

This point of booking immediately and planning well in advance can never be stressed enough, particularly if you are aiming to view a total solar eclipse. Some enthusiastic eclipse watchers book their places along the path of totality many months and often a year or two before the date of the event itself (often because they have already experienced being at the back of the queue on some previous occasion and don't want to go there again).

When you are making your travel and hotel or guest house arrangements, don't overlook things like booking your pets into suitable accommodation at the same time. There is nothing worse than having reserved your own accommodation and made your travel arrangements only to have to try and find someplace to take your cat or dog at the last minute because the boarding kennels were booked solid a couple of months previous.

Of course, it may be that your chosen viewing site is so close that you can do without arranging any special accommodation. If this is the case then all well and good, but be prepared to get caught out once in a while if you are relying on your own transportation. The most spectacular (and even many lesser) eclipse phenomenon draw huge crowds and the resulting gridlock on the roads can easily double the amount of time it takes to get to and from your destination. For this reason you may want to consider staying in the area anyway for an extra day before and after the event so that you can try and avoid such congestion.

Supplies and Equipment

The supplies and equipment you will need to make your viewing of the eclipse most beneficial and pleasurable will naturally depend on how far you intend to travel, for how long and whether or not you intend to record the eclipse in any way. Here are some items which would be handy for any eclipse viewer to take with him, a few of which will be discussed in more detail in later chapters.

Camera

Even if you don't intend to take photographs of the eclipse, take a camera with you anyway. The number of people who regret not having a camera handy during their first eclipse is extraordinary. By packing a camera and ensuring you have it with you on the day you will at least be confident that you could change you mind and take a couple of pictures if you get the urge to do so. If you intend to photograph the eclipse, full details of how to do this more effectively can be found in Chapter Six.

Emergency supplies

If you are driving in your own vehicle to view the eclipse, make sure that you have plenty of emergency supplies with you. As we said earlier, the roads can become very crowded when tens or even hundreds of thousands of people converge on one fairly small area, and the result can often be gridlock. Having a mechanical breakdown or a minor accident in such circumstances can be particularly distressing, so have a blanket, a few chocolate bars, a local map and a good book with you just in case the worst happens.

A raincoat

Regardless of what the weather forecast is for your viewing site, make sure you take a raincoat with you. Weather forecasts can be wrong, and if there is a sudden shower before or after the event you don't want to let that spoil things for you. Having a small pocket raincoat with you will help keep your clothes dry and your spirits high should the forecast go awry.

Viewing equipment

Here we are talking about packing a telescope, a pair of binoculars, a pair of special eclipse glasses or a small home-made projector, depending on what kind of eclipse you will be viewing and what kind of equipment (if any) you would prefer to use. Because viewing equipment can have such an important part to play in enhancing your eclipse experience, we will discuss it in full detail in Chapter Five.

A sheet

If you are going to be viewing a total solar eclipse, you might find it useful to take a white sheet with you. This can be laid on the ground before the event begins and will make it a great deal easier to observe any shadow bands which may appear.

Tape recorder

If you are going to record the eclipse with a camera, you will naturally miss out on the great sense of atmosphere which is experienced during major eclipse events where lots of people are present. One way to capture this atmosphere is to take a tape recorder to the site with you and to record the sounds of the event from start to finish. Many enthusiastic eclipse viewers like to do this as a matter of course - even if they are video taping the event - because they like to have a pure sound recording which they can listen to and jog their memories at a moment's notice. One thing is for certain, the first time you play back such a tape, you will be surprised at how "scripted" things sound - especially at the moment when the Sun enters totality and everyone gasps at the same time.

It would be a good idea to record the sounds of the event even if you don't currently have any desire to listen to it at a later date. After all, your children or grandchildren may take great pleasure in looking through your photographs and listening to the sound-track at some point in the future.

Sketching materials

Taking video footage or photographs is the modern way to record an eclipse, but the art of sketching an eclipse as it happens is something which many people continue to do as a hobby. It isn't as difficult as you might think, and all you will need are a few coloured pencils and a small sketch pad. If you decide to record the eclipse event in this way, be sure to read the sketching instructions in Chapter Seven.

Waterproof mat

If you are to be viewing a total solar eclipse, you will probably already realise that the amount of time between first contact and second contact can be more than many people think - in many cases more than an hour will elapse between the two stages. If you have great legs and enjoy standing even when little seems to be happening then that's great, but if you'd rather have the option of taking a seat, a waterproof mat (rubber car-mats are popular for this purpose) will help to protect you from cold and, possibly, wet ground.

Local map

No-one likes to admit that their sense of direction is anything less than it ought to be, but the sad fact is that many people fail to reach their chosen viewing sights in time to get a good eclipse experience simply because they lose their way. Often this is no real fault of their own - it's just that tens of thousands of people converging on one relatively small area can have a very disorienting effect. Getting lost, therefore, is relatively easy.

A small local map which covers your viewing site and the s urrounding areas is an essential tool which, hopefully, you will never need. Murphy's Law, of course, states that if you don't have one you will most definitely need one, so pack the thing just in case.

Thermos

If you are going to be viewing a total eclipse in winter or a lunar eclipse on a cold night, you should make sure you pack a thermos of hot liquid such as tea or coffee. Often the spectacle of an eclipse in the happening can make us lose our awareness of just how cold the weather is. Having a thermos of hot coffee handy will help ensure that you can warm yourself up to some extent without needing to leave the viewing site and potentially miss out on a portion of the experience.

Accessories

If you are going to take a tape recorder, make sure that you also take at least one set of spare batteries and a spare cassette tape. If you are taking a video camera, hand-held camera, telescope or binoculars you may also need spare batteries, videos, films and perhaps a special lens cap or two (see Chapter Six).

The rule here is to consider the accessories you may need to take with you to ensure that other items you are taking function at their best. It's no good taking a video-camera to record the eclipse if you haven't got a spare battery or video cassette. Plan ahead now and be ready for anything.

You should remember that not all of these supplies will be needed for every eclipse you set out to view. The items have been mentioned because most or all would be useful on a longer trip to view a particularly important eclipse. Obviously, if you are going to be viewing a lunar eclipse in your own back garden you can safely overlook items such as a local map, emergency supplies and so on.

Whatever type of eclipse you intend to view, do at least consider the items we have discussed in this chapter. Merely reading through the description of items may help ensure that you don't forget something you particularly want to have at hand, but a quicker and better way would be to consult the checklist which follows over the page and tick off each item you feel you will need on the day of the eclipse.

When you have done this you can go through your list and tick the relevant items again when you have packed them. This simple system will ensure that you are as fully equipped as possible and won't have to have a second-rate eclipse experience due to simple (and all too common) absent mindedness.

Supplies & Equipment Checklist

	Need	Packed
Sketching materials		
Coloured pencils	[ ]	[ ]
Sketch pad	[ ]	[ ]
Thermos	[ ]	[ ]
Viewing equipment		
Telescope	[ ]	[ ]
Lens cover	[ ]	[ ]
Tripod	[ ]	[ ]
Binoculars	[ ]	[ ]
Lens covers	[ ]	[ ]
Tripod	[ ]	[ ]
Home-made projector	[ ]	[ ]
Eclipse glasses	[ ]	[ ]
Miscellaneous items		
Waterproof mat	[ ]	[ ]
Raincoat	[ ]	[ ]
Local Map	[ ]	[ ]
A sheet	[ ]	[ ]
Emergency supplies		
Chocolate bars	[ ]	[ ]
Blanket	[ ]	[ ]
Good book	[ ]	[ ]
Recording equipment		
Tape recorder	[ ]	[ ]
Batteries	[ ]	[ ]
Cassette tapes	[ ]	[ ]
Camera/Camcorder	[ ]	[ ]
Lens cover	[ ]	[ ]
Batteries	[ ]	[ ]
Film	[ ]	[ ]
Tripod	[ ]	[ ]

Chapter Four

Viewing the Eclipse

Viewing the eclipse you have prepared for is not always as simple as looking up to the skies and watching. Of course, it can be this simple. If you are intending to observe a lunar eclipse from your own back yard then viewing can often be as easy as stepping outside and looking up. However, even in this kind of situation you may want to view the spectacle more closely, and that is where you should start to think about using a telescope, a pair of binoculars or some other viewing aid.

Because a lunar eclipse differs quite dramatically from its solar counterpart, we will take a detailed look at how to view each later on in this chapter. First, however let us spend some time discussing the different eclipse viewing options available to the budding amateur astronomer...

The Telescope

The telescope is the most popular viewing aid among both amateur and professional astronomers. Invented in the Netherlands in 1608, the most common telescopes in the early years were based on instruments used by the famous scientist Galileo, which consisted of two simple lenses fixed in a tube.

Astronomers come a long way since those early days and modern telescopes are much more powerful and accurate as far as image size and clarity is con-

The refracting telescope is the modern day version of the early telescopes used by Galileo. It works by using an objective convex lens made from glass to collect light and focus it down the tube of the instrument to the eyepiece.

cerned. In addition, there are a number of different types of telescope available, the main ones being the reflector and the refractor.

The reflecting telescope was first invented by Sir Isaac Newton and it works by using a concave mirror to collect and focus light. Because high-quality mirrors are easier and more economical to produce than lenses, virtually all large telescopes are reflectors.

As far as the eclipse enthusiast is concerned, either type of telescope can be of enormous value. Of course, the more you spend on any optical instrument, the more you can expect to benefit, but even relatively modest instruments beat the naked eye hands down when you are trying to observe the finer details of an eclipse.

Most experienced eclipse viewers agree that when photographing an eclipse, using a telescope to magnify the image produces a more satisfactory result than using one side of a pair of binoculars for the same purpose. You should, however, bear in mind that some telescopes can be quite cumbersome. For this reason, it makes sense that if you intend to travel regularly to eclipse viewing sites you should give at least some thought to portability. Modern compact telescopes which are readily available from any specialist optical store therefore present themselves as ideal viewing instruments as far as eclipses are concerned.

One additional purchase you should consider making when you obtain a telescope is a suitable tripod. Mounting the instrument on a tripod will ensure that the image you enjoy is free from shakes and wobbles, and is absolutely essential if you intend to photograph the phenomenon through a telescope.

We will be discussing the use of a telescope to view eclipses a little later in this chapter, but it would be prudent for me to point out here that one should never observe the Sun through a telescope without taking proper safety precautions. To observe an eclipse safely you need to ensure that your telescope is fitted with a solar filter. These are usually available from the store you buy the telescope from, so do ask about these when if and when you make your purchase.

Binoculars

Binoculars, like the telescope, were also invented in 1608 - this time by a Dutchman named Hans Lippershey. They are effectively two telescopes which are moulded side by side to provide the user with a stereoscopic view of the magnified image. Their compact size in comparison to telescopes is achieved through the use of prisms which can be said to "fold" the light path from the objective lens to the eyepiece.

Binoculars are becoming increasing popular with eclipse viewers for a number of good reasons. The first is their portability. Whilst lugging a large telescope to the other end of the country, or even to another country altogether, is often impractical, taking a pair of binoculars with you poses no problem at all. Another advantage of binoculars is that they require no setting up, although if you choose to do so you can mount them on a suitable tripod in order to eliminate any shakes or wobbles you may otherwise experience.

The most preferred binoculars among eclipse enthusiasts have fairly wide fields of vision, such as 7 x 35 or perhaps 7 x 50. Larger binoculars can also be used effectively, but become too cumbersome for many people to hold over long periods of time. So if big is your style, a tripod may be a wise and strain-relieving investment.

Buying an optical instrument

Whether you want to use a telescope or a pair of binoculars to view eclipses, you should not rush into making a purchase immediately. There are a vast number of optical instruments available and a wide range of prices being charged. So it would be a good idea to narrow down the options available by deciding:

What purpose do you want the instrument to serve?
Are you looking for an instrument which you can use generally, or is your sole intention to use the instrument for astronomical purposes?

How often you intend to use the instrument?
Are you looking to use the instrument just once or twice a year, or do you intend to use it on a regular basis? If the former then a relatively cheap model may be sufficient, but if the latter then you will need something a little more robust.

How much you wish to spend?
When all is said and done, your answer to this question will automatically narrow down your options more than any other factor. The bottom line is that the more you are willing to spend on a telescope or a pair of binoculars, the more choice you will have. You should bear in mind, of course, that even modestly priced instruments can be of enormous value, so don't be afraid to stick to a sensible budget.

When you have the answers to these three questions, you are ready to make a purchase. Visit a number of stores which sell telescopes and binoculars and ask for the advise of the sales personnel based on your requirements. Get a few prices. Don't however, make any on the spot decisions unless you truly feel that you have found your ideal instrument. Instead, go home (where no sales man or woman can influence you) and weigh-up the pros and cons of each. When you have done that you can purchase the item which is most suited to your particular needs and pocket.

Eclipse Glasses

Eclipse glasses are specifically designed to allow the observer to look directly at a solar eclipse without damaging his eyesight in the process. They are almost always stand-alone items and should therefore never be used in conjunction with any magnification aid such as a telescope or binoculars. They should also be avoided by people who have existing eye problems, since there is a possibility that these could be worsened by looking at the Sun despite the protection such eclipse glasses offer.

The glasses themselves come in all shapes and sizes, but typically the special lenses are fashioned from aluminised mylar. Such lenses filter the glare of the Sun down to a fairly safe level, but only if they are in pristine condition. Damaged lenses can allow too much light to seep through and thus potentially damage the eyes, so check the glasses thoroughly before use.

You can receive a FREE pair of solar eclipse viewing glasses as a reward for buying Eclipse by turning to the back page and returning the coupon. You will also find details of how to obtain extra sets of glasses.

Viewing a Lunar Eclipse

Because watching a lunar eclipse does not involve looking at the Sun in any way, shape or form, the event can be enjoyed directly through a telescope or a pair of binoculars without you having to worry about taking any particular safety precautions or using special lenses. You can, of course, simply observe the event with the naked eye, if that is what you prefer.

However you choose to view the eclipse, it is important to prepare yourself at least an hour or two in advance of the spectacle even if you are only going to be viewing from your own back yard. Doing this will ensure that you get a clear line of sight and do not discover at the last pre-eclipse moment that a building is blocking your view.

Once you have set up, the event itself can be viewed quite naturally. Viewing a lunar eclipse is delightful in its simplicity - you just observe and take the spectacle in. Of course, if you wish to record the eclipse then other steps need to be taken, but these will be covered in Chapter Seven.

You don't have to be exactly on the central line to view a total eclipse.

Viewing a Solar Eclipse

The power of the Sun is enormous and should never be underestimated. As we stated earlier in this chapter, under no circumstances should you look directly at the Sun through a telescope or binoculars without first ensuring that you have first fitted a suitable solar filter which eliminates any potential damage that can be done to your eyes.

As you might imagine, viewing a solar eclipse is more involved than viewing its lunar counterpart. For a start, you will often need to travel many miles in order to get the best viewing position. But even if you can view a solar eclipse from your own neighbourhood, there arc several preparations you may need to make.

For a start, you need decide exactly how you want to view the eclipse. There are four main options:

- You can view the eclipse through a telescope or a pair of binoculars provided that a suitable filter has been fitted. The advantage of this method is that you can get a magnificent close-up image of the eclipse, and this will make the event one which you are likely to remember for the rest of your life.

- You can view the eclipse using the eclipse glasses given free with this book. This is a very economical and relatively safe way to view the eclipse without using any magnification instruments.

- You can view the eclipse by building a pin-hole projector of some kind. This is most suitable for group viewing, as several people can witness the eclipse simultaneously without arguing over one optical instrument. Details of how to build projectors are given in the next chapter.

- You can watch the event on television. This doesn't sound as exciting as any of the other options, but in many cases it is one of the best. Eclipses broadcast on TV are often photographed from above cloud level (in many cases from outside the Earth's atmosphere), ensuring that even the bleakest of weather conditions do not spoil the event. Another advantage is that the images broadcast on television are of a much higher quality and magnification than an amateur observer could ever reasonably expect to enjoy using modest optical instruments.

Watching the eclipse through eclipse glasses, via a pin-hole projector or on television are all very straightforward methods in as much as all you need to do is observe. Using a magnification instrument such as a telescope or a pair of binoculars, however, warrants further discussion.

First of all, you must ensure that a suitable solar filter is fitted so that your eyes will not be damaged when you view partial phases of the eclipse. I realise that I have said this a number of times already, but safety is such an important issue in solar eclipse servation that the point simply cannot be stressed enough. Many first time observers hold the opinion that the chances of their eyes being damaged during a partial phase are less than they are when the Sun is unobscured. This is simply not true. The only relatively safe time to observe an eclipse without a solar filter is during totality itself, but even then you need to be very careful. The fact is that the moment totality gives way to partiality, the suns rays peek out from behind the Sun and, if you are viewing the event through a telescope or binoculars at this point without a filter, you could well find that your eyes are irrevocably damaged.

Once a filter has been fitted, you need to set up your instrument in such a way that you will be able to enjoy a clear view of the eclipse and be comfortable at the same time. How easy this is to achieve depends on a number of factors, namely the position of the Sun in the sky, your environment and whether or not you are going to mount your telescope or binoculars on a tripod. Generally speaking, if the Sun is going to be fairly high in the sky at the time of eclipse it should not be too difficult to position yourself that you can observe from a seated position, either on the floor or on a small deck chair or similar. If the Sun is going to be lower then you will have to ensure that your are high enough to observe the event, taking into account any people who may be in front of you and any buildings or trees in the environment which could otherwise obscure your view.

The final preparatory step is to lay out your white sheet on the ground so that you can view the shadow bands at the appropriate time. It is, of course, possible to view shadow bands without doing this, but using a white sheet will make the bands far easier to observe and, if you are so inclined, to photograph or film.

When everything is set up, all you need to do is sit back and wait for the eclipse to take place. Then you can observe and enjoy the spectacle without having to make any last minute adjustments.

Chapter Five

Viewing For Children

Most children enjoy the spectacle of an eclipse, but because younger children do not automatically understand the nature of an eclipse, or the dangers involved in looking at a solar event through a telescope or binoculars, they should be supervised by an adult at all times.

In order to help children get the most out of the eclipse, it is helpful to explain to them a few days beforehand how the phenomenon occurs. This explanation needn't be like a science lesson, but can be illustrated through the building of simple models.

Explaining the solar system

The best way to start educating children is to explain our place in the solar system. This can, of course, be achieved by illustrating the planets in relation to the Sun on a blackboard or sketch-pad. But, perhaps, the most effective method is to build a small model of the solar system using a few household items. This allows the children themselves to get involved in the educational process, and makes learning fun.

To build a simple model of the solar system (which will not be to scale but will be useful nevertheless), you will need a tennis ball and nine ping pong balls (or an apple and nine tangerines). You will also need a marker.

✔ *Begin building your solar system by placing the tennis ball on the floor. Make sure there is plenty of room all around as this represents the Sun - the centre of our solar system.*

✔ *Next, take the nine ping pong balls and get the children to write the name of a planet on each. To refresh your memory, the nine planets are: Mercury, Venus, Earth, Mars, Jupiter, Saturn, Uranus, Neptune and Pluto.*

✔ *Once all nine ping-pong balls have been marked, get your children to place them "in orbit" around the Sun. The position of the planets in relation to the Sun as illustrated.*

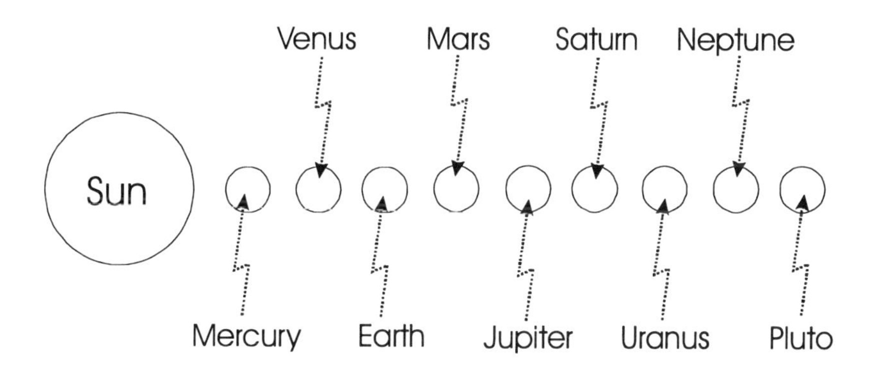

Figure 6

Once the planets have been set out in this way, you can clearly illustrate how the Earth orbits the Sun (or any of the other planets for that matter) by rolling the relevant ping-pong ball around the tennis ball in an oval path. Of course, it is impossible to move all nine ping pong balls simultaneously, so the principle of relative orbits can at least be demonstrated on a basic level.

Explaining an eclipse

When a child understands how planets orbit the Sun, he or she is only a stones throw away from understanding the eclipse phenomenon. Building a simple model of a solar eclipse takes just a few minutes, but can be worth hours of text-book education.

✔ Begin by setting a torch about one metre away from a wall. The light from the torch should be bright against the wall, so draw the curtains or dim the main lights if necessary.

✔ Now explain that the torch represents the Sun and the wall represents the surface of the Earth. It is clear from this that the light from the Sun has no problem in hitting the surface of the Earth because it is not being obstructed in any way.

✔ Introduce a tennis ball into the equation and explain that this represents the Moon. As the Moon passes between the Earth (the wall) and the Sun (the torch) a shadow is formed on the Earth. Illustrate this by actually moving the ball in front of the torch, simulating a solar eclipse. At this point you might also explain that from the Earth's point of view, the Sun seems to have disappeared, because the Moon is now blocking its light. If the torch is not too bright, perhaps ask the child to stand by the wall and look towards you. You may also vary the distance of the ball from the torch and show the different effects such as total and annular eclipses.

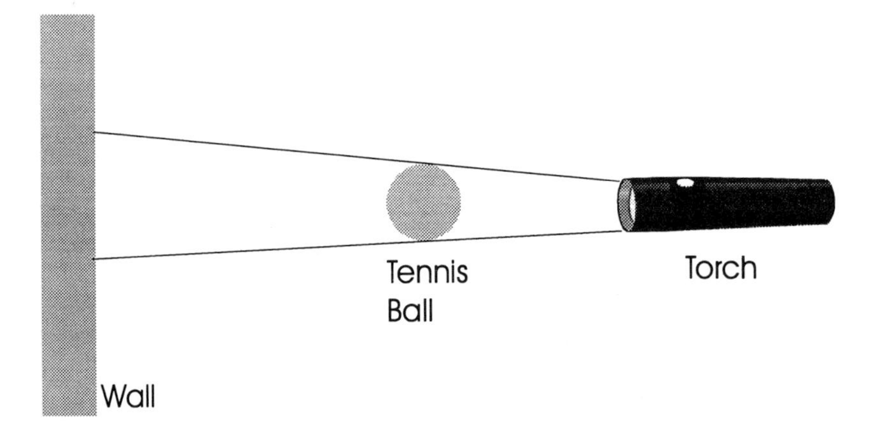

Figure 7

A lunar eclipse can be illustrated using the same simple model, but in this case you should explain that the wall represents the surface of the Moon and the tennis ball represents the Earth. As the Earth passes between the Sun and the Moon, the amount of light on the Moon decreases, giving us what we refer to as a lunar eclipse.

As we said earlier, these models are not to scale or in any way scientific, but they are useful in teaching young children how eclipses occur. This will remove any fear a child may have when seeing the Sun or the Moon "disappear" so dramatically for the first time.

Building a pin-hole projector

Having covered the topic of explaining the eclipse phenomenon to young children, we can now talk about one of the best tools you can use to help them view the actual event in complete safety - the pin-hole projector.

As the name implies, a pin-hole projector projects an image (albeit inverse) of the Sun onto a screen of some sort so that the children can watch the eclipse without actually looking at the Sun itself. Indeed, the results given by the use of pin-hole projectors are so good that many adult eclipse viewers use them instead of any other method. An adult version of this projector will be detailed later in this chapter.

To build the projector, you will need a large cardboard tube (the kind used send posters through the mail are ideal for this purpose), a sheet of card and a sheet of white tissue paper. Attach the sheet of card to one end of the cardboard tube and pierce a hole (just a couple of millimetres in diameter) into the centre so that light can travel down the tube. Fasten the sheet of tissue paper to the other end of the tube using sticky tape or glue.

To use the projector, simply lift the end of the tube with the hole towards the Sun. An image of the Sun will then be projected onto the tissue paper "screen" on the other end. Ensure that the children look at the tissue paper instead of the Sun itself and they will be able to enjoy the event in total safety.

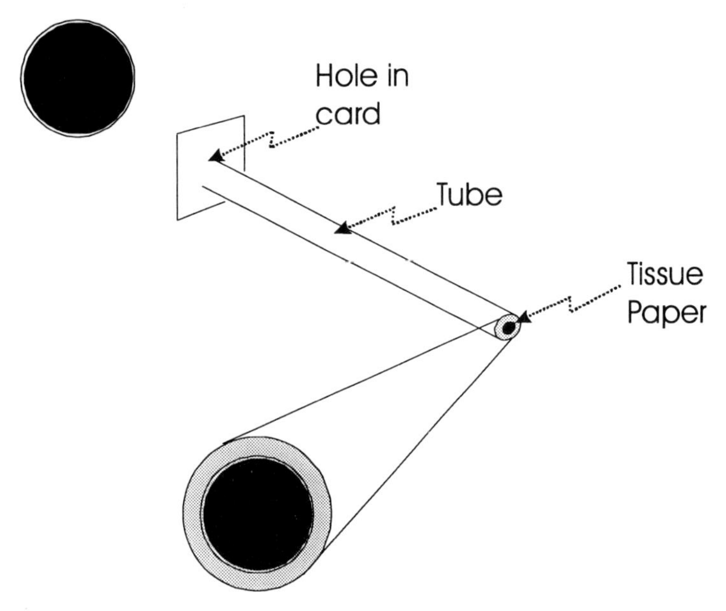

Figure 8

This simple projector is ideal for use with a very small group of viewers. For larger audiences, however, you will need to build something a little more complex and use either a telescope or one side of a pair of binoculars to provide an element of image magnification.

Building a larger projector

To build a projector, which provides a magnified image, you will obviously need a telescope or a pair of binoculars. You will also need a large sheet of white card to project the image onto, and a smaller sheet of card to shield the projected image from unwanted light.

NOTE: *Before we discuss the how-to of actually putting these components together, you should note that projecting a solar image through magnification instruments can damage the lenses of such instruments. It is therefore recommended that you do not use expensive equipment, but rather obtain a cheap model specifically for this purpose. It may surprise you, but even a child's "toy" telescope can provide an excellent projected image when used properly.*

The first thing to do is take the smaller sheet of card and cut out a circle which is the same size as the objective end of your telescope or one side of a pair of binoculars. Slip the objective end into this circle so that the card acts as a guard.

Now aim the telescope (or binoculars) towards the Sun, and place the larger sheet of white cardboard a couple of feet away from the eyepiece. If all is well, a projected image of the Sun will appear on the white card. To focus this image, change the focus on your magnification instrument or move the white card closer to or away from the eyepiece. ***Under no circumstances should you use the view-finder of a telescope to look at the Sun whilst focusing. Rely only on the projected image.***

Baily's Beads taken during the solar eclipse of February 26, 1998. NASA.

Chapter Six

Recording the Eclipse

Whenever anyone talks about recording an eclipse, there is always someone who asks, "Why bother?" The argument of such people is that no recording of an eclipse can even begin to compare to the real-life witnessing of such a spectacular event, so what is the point of recording it at all?

My answer to this question is simple. Just as a wedding photographer records images of a wedding for posterity and reference, we record eclipses for the same purposes. We don't claim that a sketch, photograph or even a video tape of an eclipse can compare with the real thing, but such recordings can and do serve as excellent reminders and souvenirs of what we have already enjoyed. They are also useful tools which can be used to educate people about the eclipse phenomenon and give others just a hint of what the experience is like. Of course, the more detailed a recording, the more useful it will be. A video recording of an eclipse will therefore capture more of the atmosphere surrounding the event than a simple sketch. All forms of eclipse recording, however, are useful, and even a simple sketch can illustrate the event better than a thousand pretty words.

In this chapter we will look at three main ways of recording an eclipse. All three methods can be employed by even a total beginner, but this is not to say that they are strictly amateur methods. The fact is that the methods we will cover - sketching, photography and videography - are used by many professional astronomers and scientists, albeit in a more complex manner. This chapter should therefore be considered as a springboard which will help you to take your first steps into the fascinating hobby of eclipse recording - a hobby which, like any other, will develop more fully as you become more involved in its pursuit.

Eclipse Sketching

One of the easiest ways of recording an eclipse is by drawing what you see. The particular method of recording eclipses appeals to many people and astronomical sketching is now considered to be a hobby in its own right. No other method allows you to put your own interpretation on the experience so directly.

The beauty of eclipse sketching is it's simplicity. You don't need an art degree to create good eclipse pictures. All you need are a few basic items and an eye for detail. The rest is just a matter of developing a good sketching technique.

Begin by obtaining the necessary items. For best results you should purchase a sketch pad which is no larger than A4 in size. This will ensure that the pad will sit neatly on your lap when not in use. A heavy pencil as opposed to a light one will make actual sketching easier and thus faster.

Before the eclipse, draw a circle in the centre of the page. This will represent the Moon at the point of totality. Then, when the moment of totality comes, simply sketch what you see. In the case of solar eclipse you will be sketching the corona of the Sun, which extends from behind the Moon in all directions. In the case of a lunar eclipse you will be sketching the appearance of the Moon itself; coloured pencils are to be preferred, for they will allow you to accurately record the reddish effect the Moon takes on when eclipsed.

When your sketches are finished, date them and either simply file them away for future reference or spray them with a protective film (protective art sprays are available from any art store) and frame them. You might not think your sketches are going to be worth framing right now, but as your skill develops you will be surprised at how much talent you can display as an astronomical artist.

Photography

Contrary to what many people believe, eclipse photography does not have to be complex or difficult. Indeed, quite good results can be had with something as simple as a Polaroid "point and click" camera, so literally anyone can at least dip their toes into the water of eclipse photography. Of course, if you want hugely magnified images of an eclipse then things do become slightly more involved.

A single-lens reflex (SLR) camera is best for amateur eclipse photography, but virtually any camera will suffice as long as you give preference to slower speed films which help to give increased picture sharpness. However, if you have a camera which has a telephoto lens then you will be able to enjoy a larger image on the final prints (bear this in mind if you intend to make a new purchase especially for eclipse work).

A tripod is absolutely essential when taking pictures of an eclipse because otherwise the "hand wobble factor" could ruin the image you are hoping to record. Use a tripod which is specifically designed for your camera and this factor will be dramatically reduced. If you want to eliminate it altogether, you will need to ensure that your camera can be operated remotely via an extended cable. Most photography specialists will be able to enlighten you on the options available (if any) for your particular instrument.

Of course, if you want to take pictures of anything other than a lunar eclipse, or a solar eclipse at the point of totality, you will need a suitable solar filter which can be placed over your camera lens. Once again, your local specialist store will be able to advise you specifically of the options available to you.

As I said a few moments ago, surprisingly good pictures can be had just be using the "point and click" method of photography, and use of a telephoto lens will enlarge the size of the image on

An Eclipse which never reached Earth! This beauty was photographed by the crew of Apollo12. NASA.

the final print. If you want to ensure that your pictures will come out as you expect them to, then it is advisable to take a few pictures of the Moon well before the scheduled eclipse. Doing this will enable you to learn how to focus your camera for best results, and will also give you some idea of the size of image you can expect of the eclipse itself.

If you find (or you already know) that you want a greater level of magnification than you can currently get out of your camera, you may wish to use a telescope to enlarge the eclipse image even further. The easiest way to do this is to use what is known as "afocal coupling". This simply means that you set up a telescope (a refractor is best for this purpose) and use your camera instead of your eye to view and record the eclipse. Coupling brackets are available from specialist photographic stores which secure your camera to the telescope itself. However, a good picture can still be taken by standing the camera tripod behind the telescope tripod and then using a black cloth to block out light between the two.

The moon makes it's move - the first photo of another Eclipse. NASA.

Again, you should practise using this afocal

coupling set-up on the Moon well in advance of the eclipse so that you can master the technique before the main event. Excellent results can be had using this method and it is therefore one of the most popular used by serious amateur eclipse photographers.

In whatever way you decide to photograph the eclipse, you should plan your shots carefully. I know from bitter experience how easy it is to shoot picture after picture when wrapped up in the excitement of the spectacle, only to run out of film as the Sun or Moon enters totality, at which point one then has to worry about changing films before the event is over. Of course, this isn't a mistake you make twice, but you needn't make it at all if your take my advice and plan in advance how many pictures you are going to take and when.

Lots of experienced eclipse viewers take along a back-up camera just in case their first one jams the film. Indeed, the more fanatical take a second camera with film already loaded for when the first camera runs out (to save time reloading) and then have back-ups for both cameras - making four in total. There's a side issue here. How many times have you known a camera to jam? Perhaps never, or once in your life. I'm convinced that the eclipse viewers obsession with jammed cameras actually causes them to jam. Every group of twenty or so watchers will produce at least one failure. Causality? Weird.

Finally, on photographic accessories, if you can get hold of one, find a right-angled viewing attachment. Lining up your camera before and after totality can be difficult when your eyes are streaming from the brightness of the Sun blasting from behind the moon, no matter how much you squint into the camera.

Storing your eclipse images is a simple matter, and the method you use will largely depend on whether you have slides or prints. I personally prefer the former, since prints can lose some of their quality in the development process - particularly if you opt for a cheap mail order deal instead of relying on a local processing

Rules of Thumb for Eclipse Photography

Film
Use slow film (ASA 100 or less) and use slide film if possible. You won't have time to change during the action.

F-Stop
Use your maximum f-stop (in the range f5.6 to f16).

Shutter speed
Use longer times during the corona and shorter times to show solar prominences.

Focus
Focus at infinity and lock it there.

Lenses
The Sun is small when viewed from here on Earth, so use a long lens (400-800mm).

Flash
Take it off, or switch it off - permanently. You'll either wreck you own images or annoy your fellow watchers.

Accessories
Tripod. A right-angle viewing attachment. Proper filters before and after totality.

Practice
Recommended. The eclipse experience is unfortunately all to short.

laboratory. Slides also have the advantage of being suitable for projection onto a screen at a later date, and such projected images are about as life-like as you can get using a simple camera.

X-Rays and Film

If you travel abroad to observe an eclipse, you should be aware of the fact that some X-ray machines in airports, etc., can damage the quality of the pictures on photographic film. There have been many cases where eclipse enthusiasts have returned from abroad and been horrified to find that their precious images were marred or completely destroyed by the x-ray machine as their baggage passed through.

The solution to this problem is to remove your films from your photographic equipment and put it in your hand baggage so that you know exactly where it is at all times. If you then have to walk through an x-ray machine you can check beforehand with the official personnel to find out if such equipment is likely to damage the films. If it is then you will usually be able to hand the film to a member of staff in order to preserve the quality of the images - something which is impossible if the film is in luggage which is being handled by a third party.

Because the photographs taken by people are so important to them, some companies manufacture special x-ray proof bags which are specifically designed to help guard against images being destroyed or impaired as they are passed through such machines. Ask your local photographic store for advice on the availability and quality of such bags. It could mean the difference between a lot of disappointment and enjoying the shots of a lifetime. A cheaper method, but not as reliable, is to wrap your exposed film in aluminium foil.

Videography

One step up from photography is videography, and this is our third method of recording an eclipse. Modern camcorders are accessible to most people and have some excellent features which make shooting video footage of an eclipse only marginally more involved than taking still photographs.

As far as the equipment itself is concerned, almost any modern instrument which can be mounted on a tripod will give good results. If you have a camcorder with a separate display panel then this will be extremely useful in helping you get the kind of footage you want, since you will be able to make minor adjustments and notice the difference without having to check through the eyepiece every few seconds.

A tripod, again, is essential if you want the optimum in picture stability. Some camcorders have built-in mechanisms which aim to reduce the "hand wobble" we all experience from time to time, but in my opinion a tripod is a lot more reliable in this respect.

One final feature which you should look for if you are intending to buy a new camcorder is a time-lapse facility. This allows you to programme the camcorder to take footage automatically ever few seconds or minutes. The advantage of this feature is that it can help you to video an eclipse from beginning to end and automatically "compress" it into a few minutes of time-lapsed video tape.

Whilst some enthusiasts may argue that using time-lapse videography removes the "natural flow" of the event, I have found that most people get bored watching a ninety minute video of an eclipse when all they really want to see are the highlights such as Baily's Beads, the Diamond Ring effect and totality. Time lapse, in my opinion, allows one to encapsulate the spectacle and make later viewing a whole lot more enjoyable. You, of course, should make your own decision on this matter as it is nothing more than a personal preference.

When videoing a solar eclipse, you should, again, use a suitable filter. The filter will need to be used throughout an annular eclipse, but in the case of a total eclipse it can be removed from the appearance of the first Diamond Ring and left off through totality until the appearance of the second Diamond Ring. You should bear in mind that totality is often very brief, so be prepared to remove the filter and then pop it back a few seconds later, depending on the length of totality you are expecting.

Once again, the best way to learn to video an eclipse is to practise before the event. Take your camera and set it up to film the Moon well in advance of the main event. Experiment with time-lapse, maximum and minimum zooms (maximum is almost always better in my experience) and any other features which your camcorder has. By doing this you can master the technicalities of videography in advance and spend more time on the actual day enjoying, rather than sweating, over the process.

Digital Storage and Editing

Before I close this chapter and leave you to put the information I have covered to good use, I should briefly mention a very recent development in image technology which, I feel, is going to dramatically change the way we record, store and edit images. I am of course referring to the digital revolution and the ability of some pieces of equipment to hold an image in digital format.

Digital cameras and camcorders store images in much the same way as a compact disc stores music. The image is broken down into a string of digital data and this can then be stored on computer disc and even manipulated and edited using a desktop computer editing package. This might not sound very thrilling to many people, but it opens up a myriad of possibilities for the eclipse enthusiast. For example, it allows one to "touch up" an image already taken to make it sharper, or to edit out parts of the image, such as a building at the foot of a photograph which is more of a nuisance than an aesthetic feature.

As the world of computer technology progresses, digital storage and retrieval of images will become more and more popular with the mainstream public. A final word of advice to those who intend to purchase new equipment is therefore to think about purchasing digital equipment if you can afford to do so. This will give you a head-start on most other people and will ensure that your beloved images can be enjoyed in a modern format for many years to come.

A Word on Light Pollution

However you decide to record your next eclipse, do give some thought to the amount of light pollution you might experience. Bright artificial light from street lamps, neon signs and so and can, and do, ruin photographs and video footage - so be particularly careful when choosing a viewing site. I have found that the best environment in which to take eclipse photographs and video is one which is completely natural - a field in the middle of nowhere, for example. Such sites might be more difficult to find and get to, but the extra effort almost always shows up in a better quality photographic or video recording.

The Eclipse on November 3rd 1994, as photographed from Putre, Chile. NASA.

Chapter Seven

Eclipse Chasing and Amateur Astronomy:
A lifetime's enjoyment

The first time you witness a real-life eclipse, you will be hooked for the rest of your life. Eclipse watching is a very addictive pastime, and all enthusiasts come to the end of one event wondering, "when will I get my next fix"? The next chapter answers that question in great detail, but there are alternatives to just waiting for the next solar or lunar eclipse to take place in your neighbourhood. You could decide to become an eclipse chaser or you could take up amateur astronomy as a fascinating hobby in its own right.

Eclipse Chasing

Most people are content to view eclipses as and when they are accessible within their own locality or within a few hundred miles. Some people, however, take their enthusiasm for eclipses to a whole new level. So enraptured are they by the event that they make eclipse viewing a central part of their lives. They live and breath for the next event, and endlessly plan towards getting another "fix". Such people are commonly referred to as *Eclipse Chasers.*

Just as the more common "twister chasers" devote the majority of their spare time to tracking down and viewing tornadoes, so eclipse chasers devote every spare moment to studying and viewing the eclipse phenomenon. Often these people work in groups, and typically they spend most of their non-working hours either planning their next chase with strategic precision or actually on the road, enjoying the chase itself.

Whilst it has to be said that the majority of eclipse chasers are regular people like you or me who happen to chase eclipses as a hobby, a few freely admit that their enthusiasm became something of an addiction. One such person is Bob R. of Fresno, California. He relates his story like this:

"For the most part I lived a pretty average life, but I guess I've always had what you might call something of an addictive personality. I'm the kind of guy who finds something he likes and then goes all out to get more of it, like when I went from my first Marlboro to a pack a day in less than a week. And that was in high school.

"The first time I saw an eclipse, I was like... well, just blown away. It was only a partial eclipse of the Sun, but man, that thing touched me. I'm not the kind of guy who's into mysticism or anything, but when I saw that I just went 'Woah!'. It was pretty far out.

"Back then I was married to Helen, who I'd met in college. She was with me that first time, and although she said she thought the eclipse was beautiful and all, she wasn't as affected by it as I was. In fact, she went right back to her regular life. I, of course, couldn't do that. No way. The eclipse, you might say, changed my life.

"I started studying everything I could find about eclipses in the library. I worked as a salesman for an office equipment company, so my days were pretty flexible. I'd spend a couple of hours every weekday just reading stuff about eclipses - what they were, how they took place, all that kind of stuff. Then, on weekends, I'd spend the whole day studying the subject. I watched video tapes

of events and started finding out when and where the next eclipse was scheduled to take place.

"Within a six month period, I was totally hooked. I'd gotten together with a few other enthusiasts I'd met through my studies at the library and pretty soon we were all driving around the country together in a fairly old campervan, just chasing whatever kind of eclipse we could - most lunar back then.

"Then we decided to get real serious and start travelling overseas just to get a glimpse of the latest total solar, and it was the beginning of a real good time. Unfortunately, my relationship with Helen suffered simply because we had now started going in different directions. Helen wanted to spend her time on her career and I wanted to spend all my time chasing eclipses.

"Well, to cut one heck of a long story short, I ended up having to make a choice between my addiction for eclipses and my old love of Helen and my job. Maybe I was selfish, but I chose eclipses. I have another job now which allows me to take extended breaks so that I can chase with my buddies, but that's really just a way of paying for the chases themselves.

"It's funny, but a lot of folks reckon I made the wrong choice. They look at Helen and at my old job and wonder what I must've been thinking to trade all that in for a life which is pretty much on the road. I can see their point, but all I say to those people is that I'm happier now than I've ever been. I might not have time for a relationship or for a big career, but I get to spend most every free moment doing what I love. And let me tell you, there's not many people that can say that."

Obviously, Bob isn't your regular run-of-the-mill eclipse chaser. Most chasers manage to balance their enthusiasm for eclipses with their relationships and careers. They spend a few thousand pounds or dollars each year on fares and equipment, then travel as far as their vacation time and budgets will allow. A few like Bob, however, are so addicted to the events that they willingly sacrifice everything else in their lives for their obsession. Whilst that is a decision which each individual has the right to make for himself, I am of the opinion that the vast majority have got it right in viewing eclipse chasing as a hobby and no more.

Chasing eclipses can be done alone or with a group of other like-minded enthusiasts and is really a simple matter of deciding how much time, money and effort you are prepared to put into the hobby. Obviously, Bob's story is warning enough of what can happen if you allow eclipse chasing to take away from other important areas of your life, so as long as you strive for balance between chasing and your other commitments, you won't allow your interest to become an unhealthy obsession.

Once you have decided what you can reasonably devote to eclipse chasing in terms of time, money and effort, you can decide where you can afford to travel in order to view eclipses. Many chasers set themselves a radius of a few hundred or a few thousand miles from their home base and simply travel to view the eclipses which are scheduled to take place within this geographical range. Others prefer to set themselves a limit as to how many eclipses they intend to view in a given period of time - say in the next twelve to thirty-six months. These people can then find out what eclipses are scheduled to take place in this period and choose which of them they would particularly like to see.

Obviously, eclipse chasing is not for everyone. It involves a lot of travel, organisation and disruption to personal schedules, which means that many people - enthusiastic as they may be about the eclipse phenomenon - decide against chasing as a hobby. Instead, many of these decide to take up another kind of hobby which, although not specifically eclipse-related, is still incredibly interesting and fulfilling...

Amateur Astronomy

Astronomy has been called the oldest recorded science. It is the study of stars, planets, moons, meteors, comets and everything else in outer space. Unlike most sciences, however, you don't need a doctorate to enjoy it. Amateur astronomy is enjoyed by millions of people all over the globe to a lesser or greater degree, and the label encompasses everything from simple star-gazing to detailed lunar or planetary observations.

To take up amateur astronomy, all you need is the desire to observe. You can simply take a walk outside at night and look up at the sky, labelling the constellations as you see them. Or you could equip yourself with a telescope or a pair of binoculars and start studying the lunar landscape. There are scores of things you can study as an amateur astronomer, but to get you started, here are a few of the most popular...

The Moon

The Moon is the nearest celestial body to the Earth, so it follows that a great many people focus on studying it more than anything else. Even the naked eye can make out the major craters of the Moon, but armed with a magnification instrument the surface reveals itself to be an interesting place

To begin studying the Moon seriously, it would be a good idea to obtain a map or two of the lunar landscape. These will help you to identify the various features of the Moon as you see them. They will introduce you to the lunar valleys and mountains which virtually any reasonable telescope can make visible on a clear night.

Meteors

Although many people refer to meteors as "shooting stars", this is actually a misnomer. A meteor is nothing more than a streak of light which is seen when a meteoroid - a small rock - is incinerated as it passes through the Earth's atmosphere (a meteorite is called a meteoroid while travelling in space). Despite the rather unexciting explanation of the phenomenon, however, watching meteors can be incredibly enjoyable.

If you watch the sky for a long enough period on almost any clear night, you will be incredibly unlucky if you do not see at least one meteor. Some, of course, will appear as nothing more than faint traces of light that are quite unremarkable, but others literally streak across the night sky and cannot fail to be missed.

The chances of you seeing a shooting star increase dramatically if you observe the skies during periods when it is known that the Earth is entering a particularly notable "cloud" of meteoroids. The resulting astronomical display is commonly referred to as a meteor "shower" because they can be seen in great number. The main meteor showers are:

- ***The Leonids***

  This shower occurs between November 17 and 18 and is best viewed in the early morning hours as is caused by the Earth nearing the orbital path of comet Tempel-Tuttle.

- ***The Geminids***

  This shower occurs between December 12 and 14 and can be viewed at any time of night, although best viewing results can be had by those living away from sources of light pollution.

- ***The Perseids***

Probably the most famous meteor shower of them all, this one occurs between August 8 and 15. Meteors can be viewed throughout the night, but the hours before dawn are usually the most productive.

The Constellations

This is the study of the constellations, which are basically groups of stars which are collectively named due to the perceived patterns they make in the sky. Obviously, the stars in a constellation are not actually connected, it is simply that they appeared to make the shapes of animals, etc to the ancients sitting here on Earth. Indeed the distances between these are huge, especially when you consider the third dimension (i.e. some of the stars in a constellation are much further from the Earth than others). Also, many of the so-called stars in these shapes are actually galaxies.

Most people know at least a few constellations such as Orion, the Great Bear, Taurus, and so on, and anyone armed with a book of constellations can quickly learn to recognise many more of them, since one constellation can help you locate another, and so on.

The beauty of constellation study is that it can be done on any clear night. No waiting for a full Moon or for a particular meteor shower - this is something that can be enjoyed (weather permitting) on a nightly basis. You can also spot different constellations as we rotate through space during the year. And if you change hemisphere's you get a whole new bunch to view.

The Planets

A planet is defined as a large celestial body which orbits a star. Apart from our own Earth, there are eight other planets in our solar system which are more than worthy of study. Some, such as Venus and Jupiter, can be detected with the naked eye, but if you are want to seriously observe the planets in any detail then a telescope is vital. Your local astronomical supplier will be able to give advice on which kind of telescope will suit both your needs and your pocket. In general, it is accepted that the more powerful the telescope, the more useful it will be to the amateur astronomer.

To get you started in planetary study, here are a few details about each of the planets in our solar system which you will find of interest and use...

Mercury

Diameter at Equator	3,030 Miles
Distance to Sun	36,000,000 Miles
Length of 'Day'	59 Earth Days
Surface Temperature (Sun side)	467° C
Number of natural satellites	Zero

Mercury is the closest planet to our Sun and has a diameter of around 3,030 miles. Because the planet is so close to the Sun, the surface temperature of Mercury can be as much as 467° Celsius on the side facing the Sun and as low as -183°C on the dark side.

Mercury is the only planet, apart from our own Earth, which has a magnetic field. The surface of the planet resembles that of our Moon, with large craters and basins. The most famous landmark on the planet is the Caloris Basin, which is some 870 miles in width.

In Roman mythology, the planet Mercury symbolised the messenger of the Gods.

Venus

Venus is the second planet from the Sun and, being one of Earth's closest neighbours, is often clearly visible even to the naked eye. In fact, Venus can be as close as 24 million miles from Earth - closer than any other planet in our solar system. It has a diameter of some 7,500 miles and a day side surface temperature which soars to some 480° Celsius.

Diameter at Equator	7,500 Miles
Distance to Sun	67,200,000 Miles
Length of 'Day'	243 Earth Days
Surface Temperature (Sun side)	480° C
Number of natural satellites	Zero

The atmosphere of Venus is mostly Carbon Dioxide, and it is this atmosphere which traps the heat of the Sun and is therefore responsible for the high surface temperature. The surface itself has many impact craters and a number of volcanoes, some of which scientists believe could still be active.

In Roman Mythology, Venus symbolised the goddess of love and beauty.

Earth

Diameter at Equator	7,923 Miles
Distance to Sun	92,860,000 Miles
Length of 'Day'	1 Earth Days
Number of natural satellites	1

To give you some perspective on the other planets detailed here, our own Earth has a diameter of 7,923 miles and is some 92,860,000 miles from the Sun. Our atmosphere consists mainly of nitrogen (78.09%), oxygen (20.95%), argon (0.93%) and carbon dioxide (0.03%).

Mars

Mars has long been the star of science fiction and in decades gone by was viewed as an obvious source of extra terrestrial life. The reality, as recent probes to Mars have revealed, is that the surface of the planet is dusty and barren, although some scientists have claimed that there could have been some kind of basic life-form there around three billion years ago.

Mars composed of 100 Viking orbiter images. NASA.

The planet has an equatorial diameter of 4,210 miles and is approximately 141,600,000 miles from the Sun. It has two natural satellites which are known as Phobos and Deimos.

The landscape of Mars is fascinating, and has a number of features for the amateur astronomer to study, including four gigantic volcanoes near the equator, the largest of which - Olympus Mons - is fifteen miles high. Another notable feature is Valles Marineris, a valley measuring 2,500 miles long, 120 miles wide and 4 miles deep.

In Roman mythology, the planet Mars symbolised the god of war.

Diameter at Equator	4,219 Miles
Distance to Sun	141,600,000 Miles
Length of 'Day'	1.2 Earth Days
Surface Temperature (Sun side)	22° C
Number of natural satellites	2

Jupiter

Jupiter is the largest planet in our solar system and has a mass which is equal to 70% of the mass of all other planets combined. The equatorial diameter of Jupiter is a massive 88,700 miles and the planet has no less than sixteen natural satellites. The largest four of these - Io, Europa, Ganymede, and Callisto - were discovered by Galileo in 1610. Ganymede is the largest Moon in the whole solar system and is approximately the same size as the planet Mercury. In order of proximity to the surface of the planet, the names of the sixteen moons are: Metis, Adrastea, Amalthea, Thebe, Io, Europa, Ganymede, Callisto, Leda, Himalia, Lysithea, Elara, Ananke, Carme, Pasiphac and Sinope

Jupiter and two moons taken by the Voyager 1 spacecraft. NASA.

The main feature of Jupiter is something called the Great Red Spot. This is possibly a gigantic cloud of gases which measures 8,500 miles wide and 20,000 miles long. The red colour of this feature is thought to be caused by a high concentration of red phosphorous, but what is causing it is unknown.

In Roman mythology, Jupiter symbolises the king of the gods.

Diameter at Equator	88,700 Miles
Distance to Sun	484,000,000 Miles
Length of 'Day'	9 hours 51 mins
Surface Temperature (Sun side)	-130° C
Number of natural satellites	16

Saturn

The sixth planet from the Sun, Saturn is another planet which can be viewed without a telescope. It is the second largest planet in our solar system and has a complex ring system which was first observed by Galileo. These rings consist of small chunks of ice and rock which orbit the planet in many ringlet formations, giving the main rings a grooved appearance.

Saturn seen by the Voyager 2 spacecraft on July 21, 1981. The moons Rhea and Dione appear as dots to the south and southeast , respectively. NASA .

The atmosphere of Saturn consists mainly of hydrogen and helium, and powerful jet streams cause the frozen ammonia clouds to take on the appearance of swirling bands.

Saturn has an equatorial diameter of around 75,000 miles and has no less than eighteen moons. The largest of these is Titan, and the others are named Pan, Atlas, Prometheus, Pandora, Epimetheus, Janus, Mimas, Enceladus, Tethys, Telesto, Calypso, Dione, Helene, Rhea, Hyperion Lapetus and Phoebe.

In Roman mythology, Saturn symbolises the god of agriculture.

Diameter at Equator	75,000 Miles
Distance to Sun	886,100,000 Miles
Length of 'Day'	10 hours 14 mins
Number of natural satellites	18

Uranus

Uranus was first discovered by the English astronomer William Herschel in 1781. The planet has an equatorial diameter of 31,600 miles and is approximately 1.8 billion miles away from the Sun.

As well as having fifteen moons (Cordelia, Ophelia, Bianca, Cressida, Desdemona, Juliet, Portia, Rosalind, Belinda, Puck, Miranda, Ariel, Umbriel, Titania and Oberon), Uranus also has eleven thin equatorial rings which scientists believe contain long-chain hydrocarbons.

In Roman mythology, Uranus symbolises the father of the Titans.

Diameter at Equator	31,600 Miles
Distance to Sun	1,867,000,000 M
Length of 'Day'	17 hours 14 mins
Surface Temperature (Sun side)	-190° C
Number of natural satellites	15

Neptune

Neptune is the eighth planet from the Sun and has the accolade of having the highest winds in our solar system. It's main feature was known as the Great Dark Spot, similar to Jupiter's Great Red Spot. The feature was approximately the same size as our own Earth, but pictures taken by the Hubble Space Telescope in 1994 seem to indicate that the spot has now disappeared.

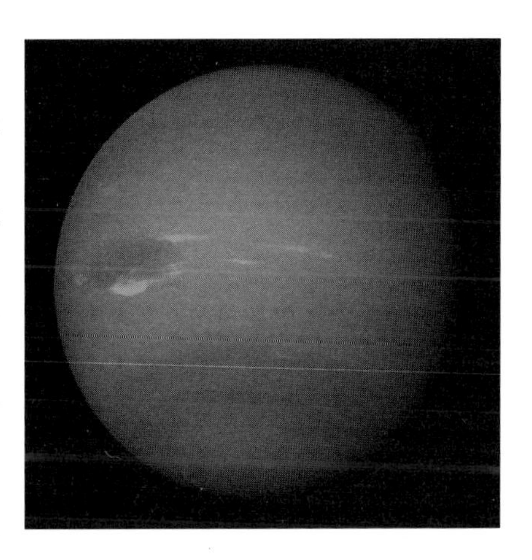

Neptune showing its "Great Dark Spot." NASA.

The planet has an equatorial diameter of 30,200 miles and is some 2.794 billion miles from the Sun. It has three faint rings and eight moons which are named Naiad, Thalassa, Deponia, Galatea, Larissa, Proteus, Triton and Nereid. Of these, Triton and Nereid can be viewed from the Earth.

Diameter at Equator	30,200 Miles
Distance to Sun	2,794,000,000 M
Length of 'Day'	16 hours 7 mins
Surface Temperature (Sun side)	-220° C
Number of natural satellites	8

Pluto

This is the smallest planet in our solar system, and because it is also currently the farthest from the Sun (being some 3.6 billion miles away) it is also the coldest. Pluto has an ice cap at the north pole and it's one Moon, Charon, is predominantly made up of ice.

The equatorial diameter of Pluto is just 1,438 miles, whilst that of Charon is 750 miles. This makes Charon the largest Moon in the solar system in relation to its parent planet.

As you might expect, the surface of Pluto is not what you would call welcoming, and is composed almost entirely of rock and ice. The atmosphere is thin and is thought to contain modest amounts of methane gas.

In Roman mythology, Pluto symbolises the god of the underworld.

Diameter at Equator	1,438 Miles
Distance to Sun	3,666,000,000 M
Length of 'Day'	6.4 days
Number of natural satellites	1

Summary

Having already established at the very beginning of this book that an eclipse is simply one celestial body obscuring another, it should not surprise you that many of the planets in our solar system which have moons also experience eclipses. Of course, eclipses of the Sun, if and when they occur, can only be viewed hypothetically from the surface of the planet in question, but this fact does help beginners to remember that solar and lunar eclipses are not events which are exclusive to our own planet.

This chapter has introduced just a few of the many aspects of amateur astronomy which are open to you, and you can of course start enjoying them this evening. If you are serious about wanting to take up amateur astronomy, however, you would do well to contact your local amateur astronomy group. These groups exist so that people of a like mind can share their interests and knowledge in a social setting, and new members are almost always welcomed with open arms.

Regardless of whether or not you decide to join such a group, I am sure that your first eclipse will not be the end of your astronomical hobby. Indeed, I think it will be just the beginning.

Chapter Eight

When and Where

Now that you know how to plan for, view and record an eclipse, all that remains is for you to find out when and where eclipses are taking place. The purpose of this extensive chapter is to provide fairly detailed data so that you can do this for solar eclipses relatively quickly and accurately - not just for this year or for the next, but for the next five decades. Eclipses up to the year 2010 are covered in more detail with descriptions and maps of where you might be able to view them - so you can book your holidays for years to come!! The following chapter presents similar data for lunar eclipses.

The data was produced by Fred Espenak of NASA/Goddard Space Flight Centre and I'd like to thank him again for allowing reproduction of his hard work. First you need the key (also created by Mr Espenak) which will enable you to use it properly. The data is presented in eleven columns, as follows:

Date and UT
The Date and Universal Time (UT) of the instant of greatest eclipse. Greatest eclipse is defined as the instant when the axis of the Moon's shadow passes closest to the Earth's centre.

Type
The eclipse type is given (P=Partial, A=Annular, T=Total or H=Hybrid (Annular-Total))

Sar
The Saros series of the solar eclipse. The Saros is the name given to a "family" of eclipses by professional astronomers.

Mag
The eclipse magnitude is defined as the fraction of the Sun's diameter obscured at greatest eclipse.

Lat & Lon
The geographic latitude and longitude of the umbra are given for greatest eclipse. This data can be used with the map on the following page so that you know exactly where the eclipse will be seen best.

Alt
The Sun's altitude, in degrees.

Path
The width of the path of totality in kilometres.

Dur
The duration of totality or annularity (minutes and seconds). For both partial and non-central umbral eclipses, the latitude and longitude correspond to the point closest to the shadow axis at greatest eclipse. The Sun's altitude is always 0° at this location.

The years in this catalogue are counted astronomically. Thus, the catalogue year 0 corresponds to 1 BC, and catalogue year -99 corresponds to 100 BC, etc. Historians do not include a year 0 in dating so the year 1 BC is followed by the year 1 AD. This is awkward for arithmetic calculations and thus, the adoption of astronomical dating and the year 0 in these catalogues. *Side Note: all those who celebrate the 'new' millenium in the year 2000 do so a year earlier than they should.*

Solar Eclipses: 1999 To 2010

Local Circumstances of Greatest Eclipse

Date	UT	Type	Sar	Mag	Lat°	Lon°	Alt°	Wid	Dur
1999 Feb 16	06:34	A	140	0.993	39.8S	93.9E	62	29	00m40s
1999 Aug 11	11:03	T	145	1.029	45.1N	24.3E	59	112	02m23s
2000 Feb 05	12:49	P	150	0.579	70.2S	134.2E	0		
2000 Jul 01	19:33	P	117	0.477	66.9S	109.5W	0		
2000 Jul 31	02:13	P	155	0.603	69.5N	59.9W	0		
2000 Dec 25	17:35	P	122	0.723	66.3N	74.1W	0		
2001 Jun 21	12:04	T	127	1.050	11.3S	2.7E	55	200	04m57s
2001 Dec 14	20:52	A	132	0.968	0.6N	130.7W	66	126	03m53s
2002 Jun 10	23:44	A	137	0.996	34.6N	178.6W	78	13	00m23s
2002 Dec 04	07:31	T	142	1.024	39.5S	59.6E	72	87	02m04s
2003 May 31	04:08	An	147	0.938	66.4N	24.7W	3	-	03m37s
2003 Nov 23	22:49	T	152	1.038	72.7S	88.4E	15	495	01m57s
2004 Apr 19	13:34	P	119	0.736	61.6S	44.3E	0		
2004 Oct 14	02:59	P	124	0.927	61.2N	153.6W	0		
2005 Apr 08	20:36	H	129	1.007	10.6S	119.0W	70	27	00m42s
2005 Oct 03	10:32	A	134	0.958	12.9N	28.7E	71	162	04m32s
2006 Mar 29	10:11	T	139	1.052	23.2N	16.7E	67	183	04m07s
2006 Sep 22	11:40	A	144	0.935	20.7S	9.1W	66	261	07m09s
2007 Mar 19	02:32	P	149	0.874	61.0N	55.4E	0		
2007 Sep 11	12:31	P	154	0.749	61.0S	90.3W	0		
2008 Feb 07	03:55	A	121	0.965	67.6S	150.5W	16	444	02m12s
2008 Aug 01	10:21	T	126	1.039	65.6N	72.3E	34	237	02m27s
2009 Jan 26	07:59	A	131	0.928	34.1S	70.3E	73	280	07m54s
2009 Jul 22	02:35	T	136	1.080	24.2N	144.1E	86	258	06m39s
2010 Jan 15	07:06	A	141	0.919	1.6N	69.3E	66	333	11m08s
2010 Jul 11	19:33	T	146	1.058	19.8S	121.9W	47	259	05m20s

Solar Eclipses: 2011 to 2020

Local Circumstances of Greatest Eclipse

Date	UT	Type	Sar	Mag	Lat°	Lon°	Alt°	Wid	Dur
2011 Jan 04	08:50	P	151	0.857	64.7N	20.8E	0		
2011 Jun 01	21:16	P	118	0.601	67.8N	46.8E	0		
2011 Jul 01	08:38	P	156	0.097	65.2S	28.6E	0		
2011 Nov 25	06:20	P	123	0.905	68.6S	82.4W	0		
2012 May 20	23:53	A	128	0.944	49.1N	176.3E	61	237	05m46s
2012 Nov 13	22:12	T	133	1.050	39.9S	161.3W	68	179	04m02s
2013 May 10	00:25	A	138	0.954	2.2N	175.5E	74	173	06m03s
2013 Nov 03	12:46	H	143	1.016	3.5N	11.7W	71	58	01m40s
2014 Apr 29	06:03	A	148	0.984	70.6S	131.3E	0		
2014 Oct 23	21:44	P	153	0.811	71.2N	97.1W	0		
2015 Mar 20	09:46	T	120	1.045	64.4N	6.6W	18	463	02m47s
2015 Sep 13	06:54	P	125	0.787	72.1S	2.3W	0		
2016 Mar 09	01:57	T	130	1.045	10.1N	148.8E	75	155	04m09s
2016 Sep 01	09:07	A	135	0.974	10.7S	37.8E	70	100	03m06s
2017 Feb 26	14:53	A	140	0.992	34.7S	31.2W	63	31	00m44s
2017 Aug 21	18:25	T	145	1.031	37.0N	87.6W	64	115	02m40s
2018 Feb 15	20:51	P	150	0.599	71.0S	0.7E	0		
2018 Jul 13	03:01	P	117	0.337	67.9S	127.5E	0		
2018 Aug 11	09:46	P	155	0.736	70.4N	174.5E	0		
2019 Jan 06	01:41	P	122	0.715	67.4N	153.6E	0		
2019 Jul 02	19:23	T	127	.046	17.4S	109.0W	50	201	04m33s
2019 Dec 26	05:17	A	132	0.970	1.0N	102.3E	66	118	03m39s
2020 Jun 21	06:40	An	137	0.994	30.5N	79.7E	83	21	00m38s
2020 Dec 14	16:13	T	142	1.025	40.3S	67.9W	73	90	02m10s

Solar Eclipses: 2021 to 2030

Local Circumstances of Greatest Eclipse

Date	UT	Type	Sar	Mag	Lat°	Lon°	Alt°	Wid	Dur
2021 Jun 10	10:42	A	147	0.943	80.8N	66.8W	23	527	03m51s
2021 Dec 04	07:33	T	152	1.037	76.8S	46.2W	17	419	01m54s
2022 Apr 30	20:41	P	119	0.639	62.1S	71.4W	0		
2022 Oct 25	11:00	P	124	0.861	61.6N	77.5E	0		
2023 Apr 20	04:17	H	129	1.013	9.6S	125.8E	67	49	01m16s
2023 Oct 14	17:59	A	134	0.952	11.4N	83.0W	68	187	05m17s
2024 Apr 08	18:17	T	139	1.057	25.3N	104.1W	70	197	04m28s
2024 Oct 02	18:45	A	144	0.933	22.0S	114.4W	69	267	07m25s
2025 Mar 29	10:47	P	149	0.936	61.1N	77.1W	0		
2025 Sep 21	19:42	P	154	0.853	60.9S	153.5E	0		
2026 Feb 17	12:12	A	121	0.963	64.7S	86.8E	12	616	02m20s
2026 Aug 12	17:46	T	126	1.039	65.2N	25.2W	26	294	02m18s
2027 Feb 06	15:59	A	131	0.928	31.3S	48.4W	73	282	07m51s
2027 Aug 02	10:06	T	136	1.079	25.5N	33.3E	82	258	06m23s
2028 Jan 26	15:07	A	141	0.921	3.0N	51.5W	67	323	10m27s
2028 Jul 22	02:55	T	146	1.056	15.6S	126.8E	53	230	05m10s
2029 Jan 14	17:12	P	151	0.871	63.7N	114.2W	0		
2029 Jun 12	04:05	P	118	0.458	66.8N	66.1W	0		
2029 Jul 11	15:36	P	156	0.230	64.3S	85.6W	0		
2029 Dec 05	15:02	P	123	0.891	67.5S	135.7E	0		
2030 Jun 01	06:28	A	128	0.944	56.5N	80.1E	56	250	05m21s
2030 Nov 25	06:50	T	133	1.047	43.6S	71.3E	67	169	03m44s

Solar Eclipses: 2031 to 2040

Local Circumstances of Greatest Eclipse

Date	UT	Type	Sar	Mag	Lat°	Lon°	Alt°	Wid	Dur
2031 May 21	07:15	A	138	0.959	8.9N	71.8E	79	152	05m26s
2031 Nov 14	21:06	H	143	1.011	0.6S	137.6W	72	38	01m08s
2032 May 09	13:25	A	148	0.996	51.3S	7.0W	20	44	00m22s
2032 Nov 03	05:33	P	153	0.855	70.4N	132.7E	0		
2033 Mar 30	18:01	T	120	1.046	71.3N	155.7W	11	781	02m37s
2033 Sep 23	13:53	P	125	0.689	72.2S	121.2W	0		
2034 Mar 20	10:17	T	130	1.046	16.1N	22.3E	73	159	04m09s
2034 Sep 12	16:18	A	135	0.974	18.2S	72.5W	67	102	02m58s
2035 Mar 09	23:04	A	140	0.992	29.0S	154.9W	64	31	00m48s
2035 Sep 02	01:55	T	145	1.032	29.1N	158.1E	68	116	02m54s
2036 Feb 27	04:45	P	150	0.628	71.6S	131.3W	0		
2036 Jul 23	10:30	P	117	0.199	68.9S	3.6E	0		
2036 Aug 21	17:24	P	155	0.861	71.1N	47.1E	0		
2037 Jan 16	09:47	P	122	0.705	68.5N	20.9E	0		
2037 Jul 13	02:39	T	127	1.041	24.8S	139.1E	43	201	03m58s
2038 Jan 05	13:45	A	132	0.973	2.1N	25.4W	65	107	03m18s
2038 Jul 02	13:31	A	137	0.991	25.4N	21.8W	88	31	01m00s
2038 Dec 26	00:58	T	142	1.027	40.3S	164.0E	73	95	02m18s
2039 Jun 21	17:11	A	147	0.945	78.9N	102.1W	33	365	04m05s
2039 Dec 15	16:22	T	152	1.036	80.9S	172.8E	18	380	01m51s
2040 May 11	03:41	P	119	0.530	62.8S	174.6E	0		
2040 Nov 04	19:07	P	124	0.807	62.2N	53.2W	0		

Solar Eclipses: 2041 to 2050

Local Circumstances of Greatest Eclipse

Date	UT	Type	Sar	Mag	Lat°	Lon°	Alt°	Wid	Dur
2041 Apr 30	11:51	T	129	1.019	9.6S	12.3E	63	72	01m51s
2041 Oct 25	01:35	A	134	0.947	9.9N	162.9E	66	213	06m07s
2042 Apr 20	02:16	T	139	1.061	27.0N	137.4E	73	210	04m51s
2042 Oct 14	01:59	A	144	0.930	23.8S	137.9E	72	273	07m44s
2043 Apr 09	18:56	T	149	1.041	61.3N	152.0E	0		
2043 Oct 03	03:00	A	154	0.943	61.0S	35.3E	0		
2044 Feb 28	20:23	A	121	0.960	62.2S	25.7W	4	-	02m27s
2044 Aug 23	01:15	T	126	1.036	64.3N	120.4W	15	452	02m04s
2045 Feb 16	23:54	A	131	0.928	28.2S	166.1W	72	281	07m47s
2045 Aug 12	17:41	T	136	1.077	25.9N	78.4W	78	256	06m06s
2046 Feb 05	23:04	A	141	0.923	4.8N	171.3W	68	310	09m42s
2046 Aug 02	10:19	T	146	1.053	12.8S	15.3E	58	206	04m51s
2047 Jan 26	01:31	P	151	0.890	62.9N	111.8E	0		
2047 Jun 23	10:51	P	118	0.313	65.8N	177.9W	0		
2047 Jul 22	22:34	P	156	0.360	63.4S	160.2E	0		
2047 Dec 16	23:48	P	123	0.882	66.4S	6.5W	0		
2048 Jun 11	12:57	A	128	0.944	63.7N	11.4W	49	271	04m58s
2048 Dec 05	15:34	T	133	1.044	46.1S	56.3W	66	160	03m28s
2049 May 31	13:58	A	138	0.963	15.3N	29.7W	83	134	04m45s
2049 Nov 25	05:32	H	143	1.006	3.8S	95.4E	73	21	00m38s
2050 May 20	20:41	H	148	1.004	40.1S	123.6W	29	27	00m21s
2050 Nov 14	13:29	P	153	0.887	69.5N	1.2E	0		

Here are the more interesting eclipses taking place up to 2010 in a little more detail. This should help you plan your trips as a new Eclipse Chaser!

August 11, 1999

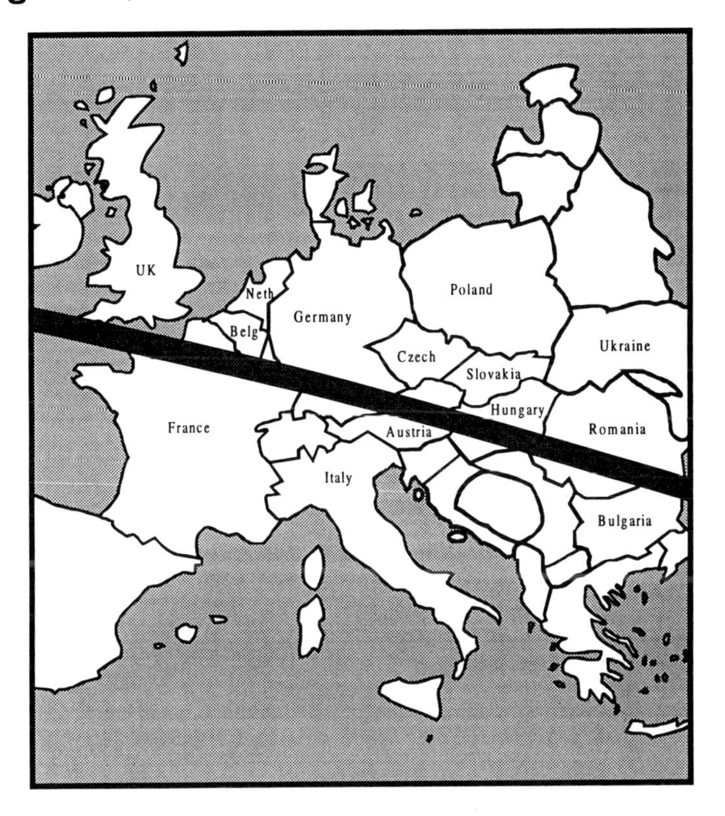

This is the last total solar eclipse of the century. The path of totality hits Cornwall in the United Kingdom, then travels across northern France and southern Germany before progressing through Austria, Hungary, Romania and Bulgaria. From there, the path hops over the Black Sea and journeys across Turkey, Iraq, Iran, Pakistan and India.

Statistically, Turkey offers the best hope of clear skies, with Romania coming in second place. The worst places as far as cloud cover is concerned are Pakistan and The United Kingdom.

June 21, 2001

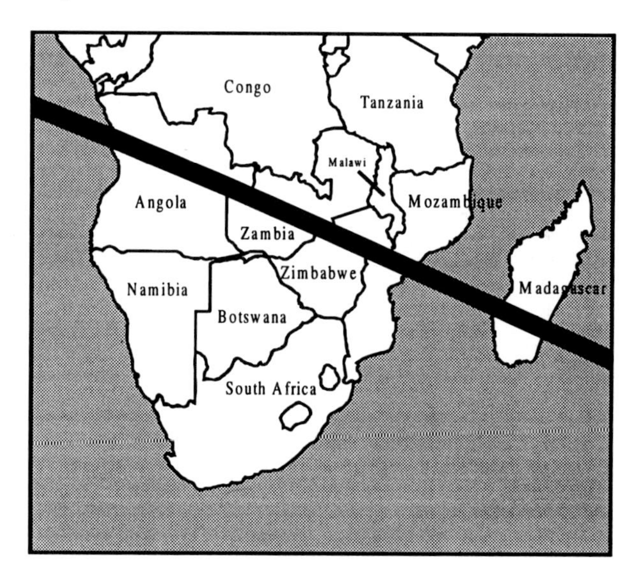

This eclipse will be very popular with those enthusiasts who have always wanted to visit Africa or Madagascar. The path of totality first hits land in Angola, then progresses through Zambia, northern Zimbabwe, Mozambique and finally Madagascar. Zimbabwe is expected to offer the best weather prospects, with Mozambique likely to be more cloudy.

December 14, 2001

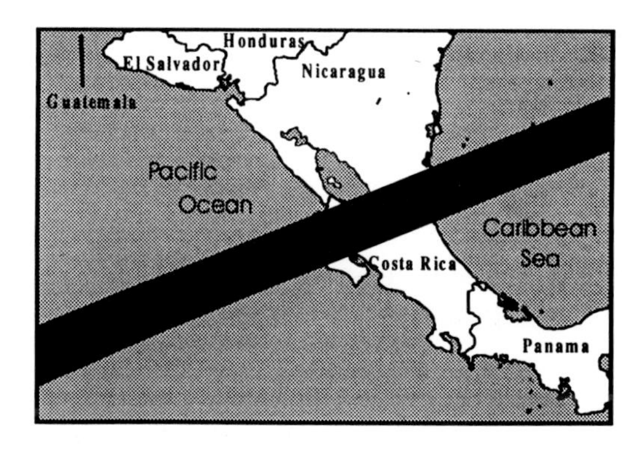

This one occurs mostly over the sea, but if you are in Costa Rica in the afternoon, you should see this annular solar eclipse.

December 04, 2002

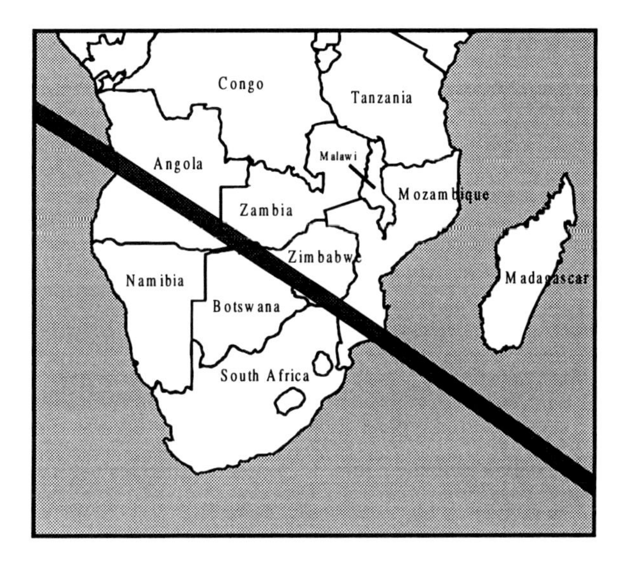

If you can't make the 2001 eclipse in Africa, this eclipse offers you a second bite of the cherry. This time, the path of totality travels through Angola, touches south-western Zambia, northern Botswana and cuts through Zimbabwe and southern Mozambique before coming to an end in Australia.

As far as weather prospects are concerned, Australia is likely to be most promising, whilst Botswana is probably going to be the least favourable, so if you ever needed an excuse to go "down under", this is one you shouldn't miss.

May 31, 2003

I have included this eclipse in this section because I find it fascinating. It is not, however, likely to be popular with eclipse chasers, simply because the eclipse will only be visible in a limited geographical area. In addition, this is not a total eclipse, but an annular event.

That said, what is it that I find so fascinating about this annular eclipse? Take a look at the map below and the answer is clear. The "path" of the eclipse in this instance forms an almost semi-

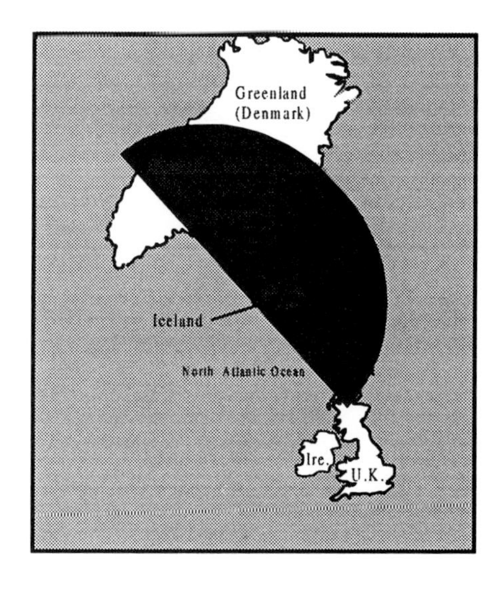

circular shape, giving much of Greenland, all of Iceland and the northern tip of Scotland a chance to witness the event.

For those die-hard enthusiasts who happen to live reasonably close to these areas, this event shouldn't be missed, though an early rise will be called for. Weather prospects are similar throughout the path of the eclipse as far as cloud cover is concerned, with perhaps the clearest skies being enjoyed by all the residents of northern Scotland.

November 23, 2003

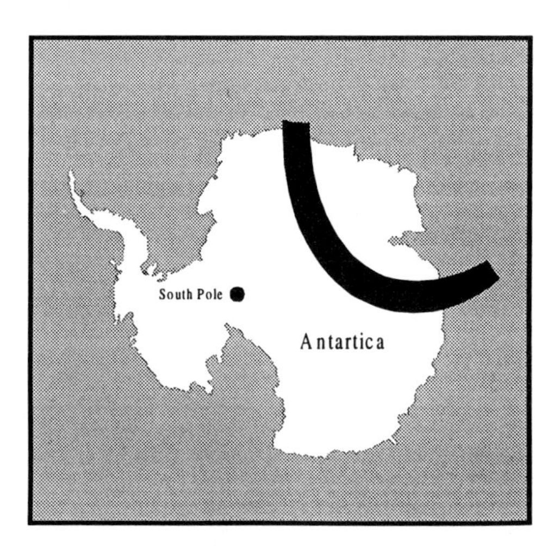

I mentioned that the total solar eclipse of May 31, 2003 is only going to be visible from a limited geographical area. Well, the same applies to the total solar eclipse of November 23, 2003... only more so.

The semi-ring shaped path of totality only ever hits Antarctica, which is not the most populated place on Earth. Even worse, totality itself will last less than two minutes, so once again I do not expect this eclipse to be at all popular with the majority of enthusiasts.

This eclipse is included to illustrate how they can sometimes be very limited geographically. Many people think that a total eclipse is visible to millions of people simultaneously if you are in the right part of the globe. This event shows that this is simply not the case. Sometimes total eclipses are visible only to fortunate few who happen to live in the right place at the right time.

April 8, 2005

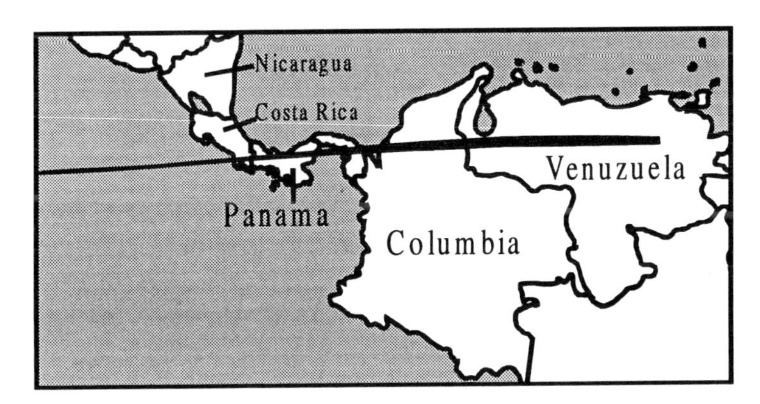

April 8, 2005 will be a date which many eclipse chasers have marked in their diary, for it is the date on which a rare hybrid (or annular-total) eclipse takes place.

As you will remember from earlier in this book, an annular-total eclipse of the Sun occurs when the Moon appears to be smaller than the Sun (as in an annular eclipse), but then appears larger as it progresses, creating a very short total solar eclipse. In this instance totality will last approximately 42 seconds, but this will take place at sea, and only an annular eclipse will be visible from land.

October 03, 2005

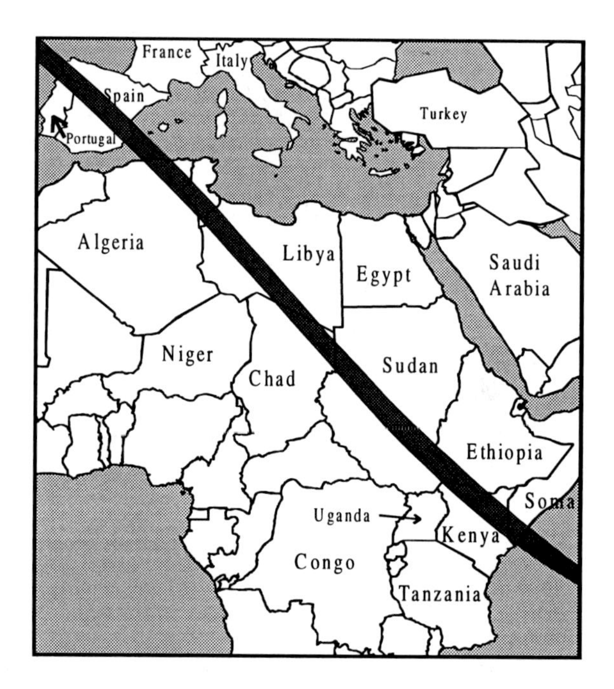

The path of this annular eclipse hits the border between Portugal and Spain, then progresses across the latter, hops over the Mediterranean Sea and then visits north-eastern Algeria, Libya, north-eastern Chad, Sudan, south western Ethiopia, Kenya and the southern tip of Somalia.

Europeans will be pleased to learn that Spain is most likely to have clear skies on the day of this eclipse, but in fact the weather prospects for all regions along the path of the eclipse are likely to be pretty good. Combine this favourable forecast with an eclipse duration of just over four and a half minutes and you have an event which no self-respecting fan of annular events will want to miss.

March 29, 2006

Yet another total eclipse visits Africa in the third month of 2006. This time, the path of totality hits Nigeria then progresses through

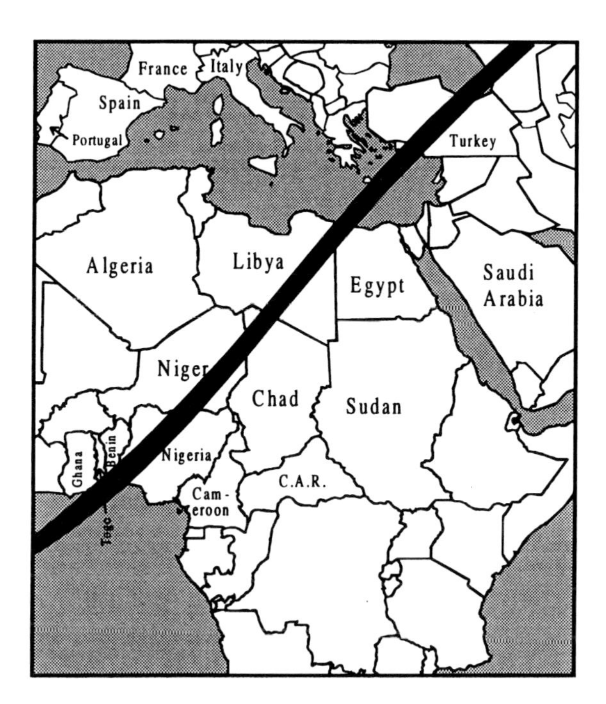

Niger, the northern tip of Chad, Libya and north-western Egypt before crossing the Mediterranean Sea and travelling through Turkey.

Although Turkey can reasonably be expected to have the clearest skies based on accepted weather statistics, my personal location for viewing this eclipse would be the eastern coast of Brazil. This is where the path of totality first hits land, and where visitors to that region will be able to enjoy a total eclipse at sunrise.

September 22, 2006

I have not provided a map for this annular eclipse simply because, despite a fairly generous path some 261 kilometres wide, witnessing the event will only be possible from Guyana, Suriname and French Guiana.

Whilst the local population in those places will no doubt enjoy the annular event, it is not one which would be of particular interest to eclipse chasers.

February 07, 2008

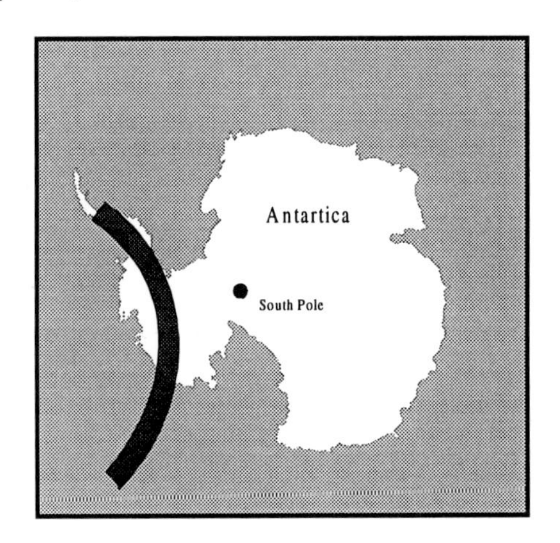

This annular eclipse is another of those which I mention only to illustrate how some eclipses can come and go almost without being noticed by the majority of the population. This particular eclipse path hits just a small portion of Antarctica, so apart from the local wildlife and residents of a nearby outpost for scientific research, the event will be missed by almost all of us.

August 01, 2008

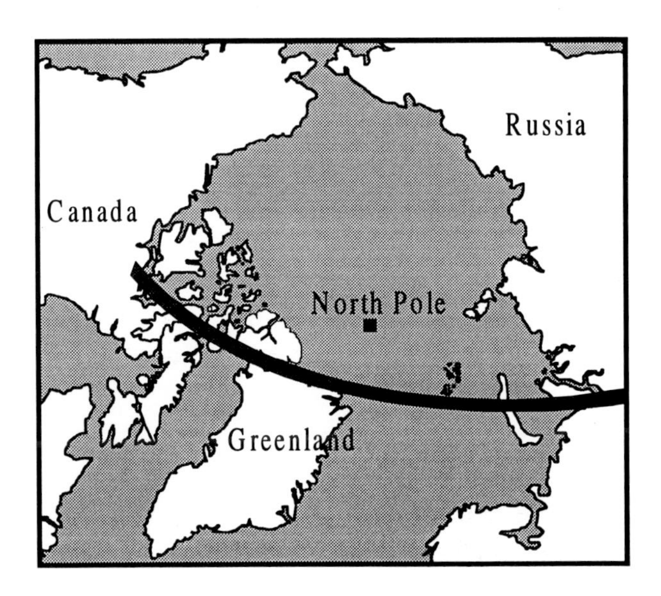

This total eclipse of August 01, 2008 is likely to be popular with eclipse chasers, being the first for over two years. The path of totality visits the edge of Canada, Greenland and then journeys into Russia, clips the western tip of Mongolia and comes to its natural conclusion in China, so some heavy-duty travelling will be involved for most of us enthusiasts. Our reward for all the inevitable jet-lag and planning? Up to two minutes and twenty-seven seconds of totality.

January 26, 2009

There is no map for this annular eclipse because, once again, visibility will be restricted to just a few people - this time, to those who live in Borneo, Sumatra or Java. Don't be fooled by the generous 7 minutes and 54 seconds of annularity either, for this will only be enjoyed by those who happen to travelling across the Indian Ocean. The good news is that to get their next total eclipse fix, hardened chasers only have to wait until...

July 22, 2009

This total eclipse will be well worth waiting for. The path of totality travels across India, completely shadows Bhutan and then

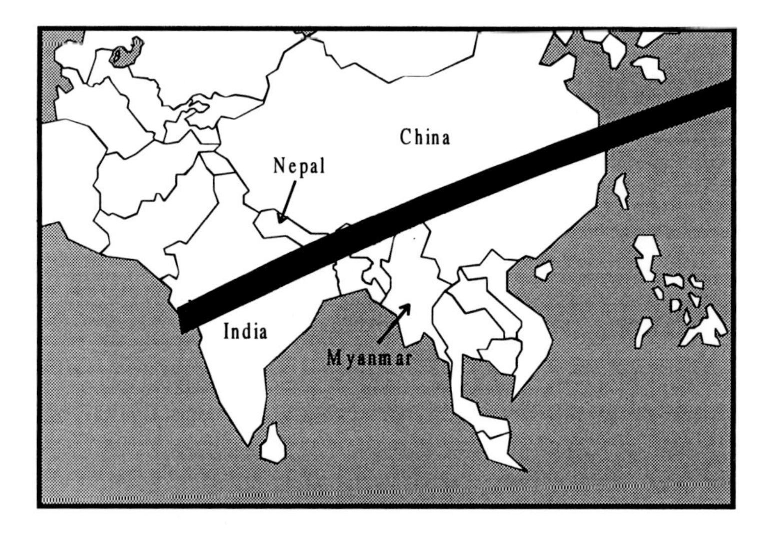

continues its journey across China. The weather prospects are likely to be best in China, and that is also where you can revel in totality for more than three and a half minutes. Indeed, in Wuhan, China, you can expect to enjoy almost five and a half minutes of totality. Definitely an event for the diary.

January 15, 2010

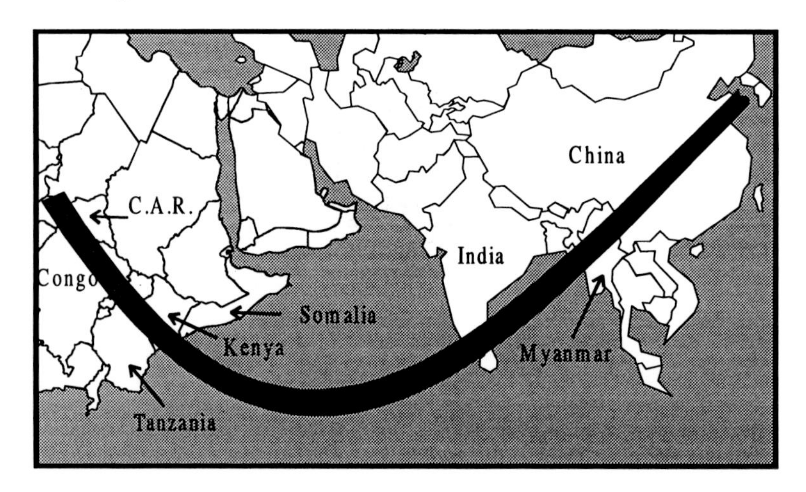

If the total eclipse of July 22, 2009 whets your appetite for lengthy events, this annular eclipse is one which you should definitely make plans to see. The path of the eclipse hits land in the Central

African Republic, travels through Zaire and Kenya, then sweeps over to clip the southern tip of India and progress up through Myanmar and China.

Wherever you view this eclipse from, you will be extremely unlucky to enjoy less than five minutes of annularity. Visit China and you can enjoy almost nine minutes - more than enough to motivate most chasers to book their tickets.

July 11, 2010

This is the last solar eclipse I will briefly mention in this book, and it is one for the die-hard enthusiast only. The path of the eclipse hits Easter Island and briefly touches Chile and Argentina, but Easter Island is likely to prove most popular with chasers, since that is where over four and a half minutes of totality can be enjoyed. Maybe this would be a good time to enjoy your summer vacation!

Chapter Nine

Some Lunar Eclipses

Just as the last chapter enabled you to find out when and where solar eclipses were to take place, so we now turn our attention to some lunar eclipses. Once again, the data contained in this chapter was produced by Fred Espenak of NASA/Goddard Space Flight Centre, as was the key.

Unlikc solar eclipses, lunar events can be seen from a much wider area of the Earth, and so no specific latitude or longitude is given in the data tables which appear in this chapter. If you desperately want to know if a lunar eclipse can be viewed from a specific geographical point, this can be calculated from the data in the table but the calculation is quite complex. In this case the best alternative would be to approach your local astronomical group and ask a more experienced astronomer to show you how to make the calculations personally based on your own latitude and longitude.

That said, the data is presented in twelve columns, as follows:

Date and UT
The date and Universal Time of the instant of greatest eclipse. Greatest eclipse is defined as the instant when the Moon passes closest to the axis of Earth's shadow. This marks the instant when the Moon is deepest in the shadow(s).

Type
The eclipse type is given (T=Total, P=Partial, or N=Penumbral)

Sar
The Saros series of the eclipse.

PMA

The penumbral magnitude of the eclipse are defined as the fractions of the Moon's diameter obscured by each shadow at greatest eclipse.

UMA

The umbral magnitude of the eclipse are defined as the fractions of the Moon's diameter obscured by each shadow at greatest eclipse.

PD

The semi-duration of the partial phase of the eclipse, given to the nearest minute.

TD

The semi-duration of the total phase of the eclipse, given to the nearest minute.

GST

The Greenwich Sidereal Time at 00:00 Universal Time.

MRA

The Moon's Geocentric Right Ascension at greatest eclipse.

MDE

The moon's Declination at greatest eclipse.

Lunar Eclipses: 1999 to 2010

Local Circumstances at Greatest Eclipse

Date	UT	Type	Sar	PMA	UMA	PD	TD	GST	MRA	MDE
1999 Jan 31	16:17	P	114	1.028	-0.021	-	-	8.7	8.91	16.4
1999 Jul 28	11:34	U	119	1.460	0.402	72m	-	20.4	20.48	-18.3
2000 Jan 21	04:43	T	124	2.331	1.330	102m	39m	8.0	8.17	19.8
2000 Jul 16	13:56	T	129	2.864	1.773	118m	54m	19.6	19.75	-21.2
2001 Jan 09	20:21	T	134	2.187	1.194	99m	31m	7.3	7.42	22.4
2001 Jul 05	14:55	P	139	1.573	0.499	80m	-	18.9	18.99	-23.4
2001 Dec 30	10:29	N	144	0.919	-0.110	-	-	6.6	6.64	24.2
2002 May 26	12:03	N	111	0.714	-0.283	-	-	16.3	16.23	-20.0
2002 Jun 24	21:27	N	49	0.235	-0.787	-	-	18.2	18.22	-24.8
2002 Nov 20	01:46	N	116	0.886	-0.222	-	-	3.9	3.71	18.7
2003 May 16	03:40	T	121	2.100	1.134	97m	26m	15.6	15.51	-18.6
2003 Nov 09	01:18	T	126	2.140	1.022	106m	12m	3.2	2.93	16.3
2004 May 04	20:30	T	131	2.288	1.309	102m	38m	14.9	14.81	-16.5
2004 Oct 28	03:04	T	136	2.390	1.313	110m	41m	2.5	2.18	13.4
2005 Apr 24	09:55	N	141	0.890	-0.138	-	-	14.2	14.11	-13.9
2005 Oct 17	12:03	P	146	1.084	0.068	29m	-	1.7	1.47	10.3
2006 Mar 14	23:47	N	113	1.056	-0.056	-	-	11.5	11.68	3.1
2006 Sep 07	18:51	P	118	1.158	0.190	46m	-	23.1	23.11	-6.7
2007 Mar 03	23:21	T	123	2.345	1.237	111m	37m	10.8	10.96	6.9
2007 Aug 28	10:37	T	128	2.478	1.481	106m	45m	22.4	22.45	-10.0
2008 Feb 21	03:26	T	133	2.171	1.111	103m	25m	10.0	10.25	10.5
2008 Aug 16	21:10	P	138	1.862	0.812	94m	-	21.7	21.76	-12.9
2009 Feb 09	14:38	N	143	0.924	-0.083	-	-	9.3	9.53	13.5
2009 Jul 07	09:38	N	110	0.182	-0.908	-	-	19.0	19.14	-23.9
2009 Aug 06	00:39	N	148	0.428	-0.662	-	-	21.0	21.05	-15.6
2009 Dec 31	19:22	P	115	1.081	0.082	31m	-	6.7	6.76	24.0
2010 Jun 26	11:38	P	120	1.603	0.542	82m	-	18.3	18.35	-24.0
2010 Dec 21	08:17	T	125	2.306	1.261	105m	37m	6.0	5.95	23.7

Lunar Eclipses: 2011 to 2020

Local Circumstances at Greatest Eclipse

Date	UT	Type	Sar	PMA	UMA	PD	TD	GST	MRA	MDE
2011 Jun 15	20:12	T	130	2.712	1.705	110m	50m	17.6	17.59	-23.2
2011 Dec 10	14:32	T	135	2.212	1.111	106m	26m	5.3	5.14	22.6
2012 Jun 04	11:03	P	140	1.343	0.376	64m	-	16.9	16.86	-21.7
2012 Nov 28	14:33	N	145	0.942	-0.183	-	-	4.5	4.33	20.5
2013 Apr 25	20:07	P	112	1.012	0.021	16m	-	14.3	14.21	-14.4
2013 May 25	04:10	N	150	0.040	-0.928	-	-	16.2	16.15	-19.4
2013 Oct 18	23:50	N	117	0.791	-0.267	-	-	1.8	1.57	11.0
2014 Apr 15	07:45	T	122	2.344	1.296	108m	39m	13.6	13.56	-10.0
2014 Oct 08	10:54	T	127	2.171	1.172	100m	30m	1.1	0.92	6.3
2015 Apr 04	12:00	T	132	2.105	1.005	105m	6m	12.8	12.89	-5.3
2015 Sep 28	02:47	T	137	2.254	1.282	100m	36m	0.4	0.29	1.5
2016 Mar 23	11:47	N	142	0.801	-0.307	-	-	12.1	12.22	-0.3
2016 Aug 18	09:42	N	109	0.017	-0.993	-	-	21.8	21.85	-11.4
2016 Sep 16	18:54	N	147	0.933	-0.058	-	-	23.7	23.67	-3.3
2017 Feb 11	00:44	N	114	1.014	-0.030	-	-	9.4	9.64	13.1
2017 Aug 07	18:20	P	119	1.315	0.251	58m	-	21.1	21.18	-15.4
2018 Jan 31	13:30	T	124	2.320	1.321	102m	38m	8.7	8.93	17.0
2018 Jul 27	20:21	T	129	2.706	1.614	118m	52m	20.4	20.47	-19.0
2019 Jan 21	05:12	T	134	2.193	1.201	99m	31m	8.0	8.21	20.3
2019 Jul 16	21:30	P	139	1.729	0.658	89m	-	19.6	19.73	-21.9
2020 Jan 10	19:10	N	144	0.921	-0.111	-	-	7.3	7.45	23.0
2020 Jun 05	19:25	N	111	0.594	-0.399	-	-	17.0	16.97	-21.5
2020 Jul 05	04:30	N	149	0.380	-0.638	-	-	18.9	18.99	-24.1
2020 Nov 30	09:42	N	116	0.855	-0.257	-	-	4.6	4.48	20.7

Lunar Eclipses: 2021 to 2030

Local Circumstances at Greatest Eclipse

Date	UT	Type	Sar	PMA	UMA	PD	TD	GST	MRA	MDE
2021 May 26	11:18	T	121	1.979	1.016	94m	9m	16.3	16.24	-20.7
2021 Nov 19	09:03	P	126	2.098	0.979	105m	-	3.9	3.67	19.2
2022 May 16	04:11	T	131	2.397	1.419	104m	43m	15.6	15.52	-19.3
2022 Nov 08	10:59	T	136	2.440	1.364	110m	43m	3.2	2.90	16.9
2023 May 05	17:23	N	141	0.989	-0.041	-	-	14.9	14.81	-17.2
2023 Oct 28	20:14	P	146	1.143	0.127	40m	-	2.5	2.16	14.1
2024 Mar 25	07:12	N	113	0.982	-0.128	-	-	12.2	12.34	-1.2
2024 Sep 18	02:44	P	118	1.062	0.091	32m	-	23.8	23.77	-2.6
2025 Mar 14	06:58	T	123	2.286	1.183	109m	33m	11.5	11.64	2.7
2025 Sep 07	18:11	T	128	2.369	1.368	105m	41m	23.1	23.11	-6.0
2026 Mar 03	11:33	T	133	2.210	1.156	104m	30m	10.8	10.94	6.4
2026 Aug 28	04:12	P	138	1.990	0.935	99m	-	22.4	22.44	-9.3
2027 Feb 20	23:12	N	143	0.952	-0.052	-	-	10.0	10.24	9.8
2027 Jul 18	16:03	N	110	0.028	-1.063	-	-	19.7	19.88	-22.3
2027 Aug 17	07:13	N	148	0.571	-0.521	-	-	21.7	21.73	-12.4
2028 Jan 12	04:13	P	115	1.072	0.072	29m	-	7.4	7.56	22.7
2028 Jul 06	18:19	P	120	1.453	0.394	71m	-	19.0	19.11	-23.3
2028 Dec 31	16:52	T	125	2.300	1.252	105m	36m	6.7	6.77	23.3
2029 Jun 26	03:22	T	130	2.852	1.849	110m	51m	18.3	18.35	-23.3
2029 Dec 20	22:41	T	135	2.227	1.122	107m	27m	6.0	5.95	23.1
2030 Jun 15	18:33	P	140	1.472	0.508	73m	-	17.6	17.61	-22.6
2030 Dec 09	22:27	N	145	0.968	-0.159	-	-	5.2	5.12	21.9

Lunar Eclipses: 2031 to 2040

Local Circumstances at Greatest Eclipse

Date	UT	Type	Sar	PMA	UMA	PD	TD	GST	MRA	MDE
2031 May 07	03:50	N	112	0.907	-0.085	-	-	15.0	14.92	-17.8
2031 Jun 05	11:44	N	150	0.154	-0.814	-	-	16.9	16.89	-21.1
2031 Oct 30	07:45	N	117	0.742	-0.315	-	-	2.6	2.27	14.8
2032 Apr 25	15:13	T	122	2.245	1.197	106m	33m	14.3	14.24	-13.8
2032 Oct 18	19:02	T	127	2.108	1.108	98m	24m	1.9	1.60	10.4
2033 Apr 14	19:12	T	132	2.197	1.099	108m	25m	13.6	13.56	-9.4
2033 Oct 08	10:55	T	137	2.331	1.355	102m	40m	1.2	0.96	5.8
2034 Apr 03	19:05	N	142	0.881	-0.223	-	-	12.8	12.88	-4.6
2034 Sep 28	02:46	P	147	1.016	0.020	16m	-	0.5	0.33	1.0
2035 Feb 22	09:04	N	114	0.991	-0.048	-	-	10.1	10.35	9.2
2035 Aug 19	01:10	P	119	1.177	0.109	39m	-	21.8	21.86	-12.0
2036 Feb 11	22:11	T	124	2.300	1.305	101m	38m	9.4	9.67	13.6
2036 Aug 07	02:51	T	129	2.553	1.459	116m	48m	21.1	21.18	-16.1
2037 Jan 31	14:00	T	134	2.205	1.213	99m	32m	8.7	8.97	17.5
2037 Jul 27	04:08	P	139	1.884	0.814	97m	-	20.3	20.46	-19.6
2038 Jan 21	03:48	N	144	0.925	-0.109	-	-	8.0	8.24	20.9
2038 Jun 17	02:43	N	111	0.467	-0.522	-	-	17.7	17.72	-22.1
2038 Jul 16	11:34	N	149	0.525	-0.490	-	-	19.6	19.74	-22.5
2038 Dec 11	17:43	N	116	0.831	-0.285	-	-	5.4	5.27	22.0
2039 Jun 06	18:53	P	121	1.852	0.891	90m	-	17.0	16.99	-22.1
2039 Nov 30	16:55	P	126	2.068	0.947	103m	-	4.6	4.45	21.3
2040 May 26	11:45	T	131	2.519	1.541	106m	46m	16.3	16.26	-21.5
2040 Nov 18	19:03	T	136	2.478	1.402	111m	44m	3.9	3.65	19.7

Lunar Eclipses: 2041 to 2050

Local Circumstances at Greatest Eclipse

Date	UT	Type	Sar	PMA	UMA	PD	TD	GST	MRA	MDE
2041 May 16	00:41	P	141	1.100	0.070	30m	-	15.6	15.53	-20.0
2041 Nov 08	04:33	P	146	1.191	0.175	46m	-	3.2	2.89	17.5
2042 Apr 05	14:28	N	113	0.894	-0.213	-	-	12.9	13.01	-5.4
2042 Sep 29	10:44	P	118	0.978	0.003	6m	-	0.6	0.43	1.6
2042 Oct 28	19:33	N	156	0.008	-0.974	-	-	2.5	2.17	14.8
2043 Mar 25	14:30	T	123	2.216	1.119	108m	27m	12.2	12.31	-1.6
2043 Sep 19	01:50	T	128	2.269	1.261	103m	36m	23.9	23.77	-1.9
2044 Mar 13	19:37	T	133	2.256	1.208	105m	34m	11.5	11.61	2.1
2044 Sep 07	11:19	T	138	2.111	1.050	103m	18m	23.1	23.10	-5.4
2045 Mar 03	07:41	N	143	0.987	-0.012	-	-	10.8	10.93	5.7
2045 Aug 27	13:53	N	148	0.708	-0.388	-	-	22.4	22.40	-8.8
2046 Jan 22	13:01	P	115	1.060	0.059	27m	-	8.1	8.35	20.5
2046 Jul 18	01:04	P	120	1.307	0.251	58m	-	19.7	19.86	-21.8
2047 Jan 12	01:24	T	125	2.291	1.239	105m	35m	7.4	7.57	22.0
2047 Jul 07	10:34	T	130	2.757	1.757	110m	51m	19.0	19.11	-22.6
2048 Jan 01	06:52	T	135	2.240	1.132	107m	28m	6.7	6.76	22.7
2048 Jun 26	02:00	P	140	1.607	0.644	80m	-	18.3	18.37	-22.6
2048 Dec 20	06:26	N	145	0.988	-0.140	-	-	6.0	5.92	22.5
2049 May 17	11:25	N	112	0.789	-0.203	-	-	15.7	15.64	-20.6
2049 Jun 15	19:12	N	150	0.276	-0.693	-	-	17.6	17.64	-21.9
2049 Nov 09	15:50	N	117	0.707	-0.350	-	-	3.3	3.00	18.2
2050 May 06	22:30	T	122	2.131	1.082	103m	22m	15.0	14.94	-17.2
2050 Oct 30	03:20	T	127	2.060	1.060	97m	18m	2.6	2.30	14.2

Conclusion

As far as natural wonders are concerned, eclipses are among the very best. There is nothing on Earth which can come close to comparing with the spectacle of one celestial body obscuring another - nothing which has the awe-inspiring power to make the hairs stand up on the back of your neck in quite the same way.

In the introduction to this book I said that my aim was to introduce you to the subject of the eclipse and give information on how you can begin to enjoy them as simply and effectively as possible. It is my sincere hope that I have achieved that aim, and that you will now become one of the many millions of people who watch eclipses with enthusiasm.

Your completion of this book is just the first step in what I hope will become a life-long study for you. The appendix which follows will enable you to take your study of eclipses even further and draw on the mass of knowledge which is available to the serious eclipse enthusiast.

I wish you a most enjoyable voyage of astronomical discovery, both now and in the future.

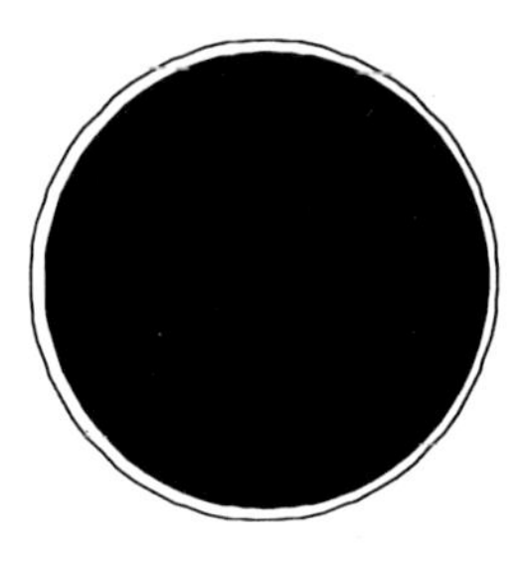

Appendix I

Further Reading

Cambridge Eclipse Photography Guide	J M Pasachoff M A Covington
Eclipse	P S Harrington
Eclipse	B Brewer
Guide to the Sun	K J H Phillips
Canon of Solar Eclipses : 1986 - 2035	Fred Espenak
The Guide to Amateur Astronomy	J Newton P Teece
Specialist Astronomical Publications	Royal Greenwich Observatory Madingley Road, Cambridge, CB3 0EZ

Glossary of Terms

Annular Eclipse
A solar eclipse where the disk of the Moon is not quite large enough to totally obscure the Sun.

Baily's Beads
The name given to the appearance of what looks like a circular string of beads, caused by light from the photosphere of the Sun peeking through the valleys of the lunar landscape.

Diamond Ring
The name given to the appearance of the Sun during a total solar eclipse when all but one shaft of light is obscured. This spectacle can usually be observed as Baily's Beads begin to disappear.

Eclipse Glasses
Specially manufactured solar eclipse glasses, the lenses of which are designed to filter out much of the glare of the Sun and thus protect the eyes of the observer.

Elliptical
Oval in shape. The term elliptical is most commonly used in reference to the shape of the orbital path of one celestial body held by the gravitational force of another.

Lunar
Pertaining to the Moon.

Moon
A natural satellite which orbits a planet.

Orbit
The elliptical path of one celestial body which is held by the gravitational force of another.

Partial Eclipse
An eclipse where one celestial body partially eclipses another.

Planet
A natural celestial body which orbits a star.

Penumbra
The lighter, outer part of a shadow.

Shadow Bands
The shadowy lines which appear to ripple on the ground during a solar eclipse. The shadow bands are caused by a thin sliver of light being twisted through the Earth's atmosphere.

Solar
Pertaining to the Sun.

Sun
A star which creates its own heat and light through a process of nuclear fusion.

Total Eclipse
An eclipse where one celestial body totally obscures another.

Umbra
The darker, inner part of a shadow.

Find What You Want on the Internet

UK£5.95/US$9.95

The sheer size of the Internet's information resources is its biggest challenge. There is no central repository of all this information, nor it is catalogued or sorted in ordered fashion.

Find What You Want on The Internet is designed to teach Internet users - from novices to veterans - how to locate information quickly and easily.

The book uses jargon-free language, combined with many illustrations, to answer such questions as:

- ❑ Which search techniques and Search Engines work best for your specific needs?
- ❑ What is the real difference between true 'search' sites and on-line directories, and how do you decide which one to use?
- ❑ How do the world's most powerful Search Engines really work?
- ❑ Are there any 'special tricks' that will help you find what you want, faster?

There is also a bonus chapter covering Intelligent Agents — special high-tech personal search programs that can be installed on your computer to search the Internet on your behalf, automatically.

Create Your Own Electronic Office

UK£5.95/US$9.95

Home-based business... Cottage industry... Small Office/Home Office (SOHO)... whatever term you use, operating from home, means you escape the stresses, pressures and overheads of a busy town centre office. What-is-more, the time saved by not having to commute will allow you to work more efficiently and spend quality time enjoying yourself.

If this sounds like the kind of independence that you have dreamed of, then this book is for you. With its help, you will:

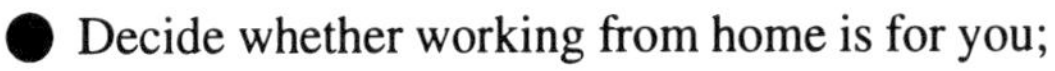

- Decide whether working from home is for you;

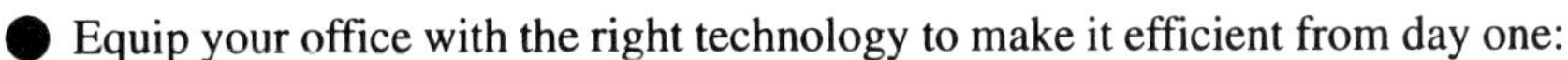

- Equip your office with the right technology to make it efficient from day one:
- Plan your new business and working environment

Included are chapters on getting yourself motivated for working by yourself for yourself, how to maintain a healthy separation between your work and private life, and how to present yourself and your new business in a professional manner.

UK£5.95/US$9.95

Complete Beginner's Guide to Word for Windows

Using Microsoft Word is a hit and miss process for a lot of people, and the end results are usually far from satisfying. What-is-more, many of the alternative books available are difficult to understand, and do not focus on the task of getting the job done, leaving you free to write creatively.

The Complete Beginner's Guide to Word for Windows is different. It has been designed and researched by the people who know best - the trainers who teach Word for a living. They understand both beginners and advanced students, and know how to meet their needs.

With clear, step-by-step instructions, and plenty of easy to understand examples, this book guides you to success the easy way. It leaves you free to concentrate on your document instead of getting the program to run properly!

Create Your Own Web Site

UK£5.95/US$9.95

The World Wide Web is being transformed into an important business and communications tool. Millions of computer users around the globe now rely on the Web as a prime source of information and entertainment.

Once you begin to explore the wonders of the Internet, it isn't long before the first pangs of desire hit – you want your own Web site.

Whether it is to showcase your business and its products, or a compilation of information about your favourite hobby or sport, creating your own Web site is very exciting indeed. But unless you're familiar with graphics programs and HTML (the "native language" of the Web), as well as how to upload files to the Internet, creating your Web page can also be very frustrating!

But it doesn't have to be that way. This book, written by an Internet consultant and graphics design specialist, will help demystify the process of creating and publishing a Web site. In it you will learn:

- What free tools are available that make producing your own Web site child's play (and where to find them);
- How to create your own dazzling graphics, using a variety of free computer graphics programs;
- Who to talk to when it comes to finding a home for your Web site (If you have an Internet account, you probably already have all that you need).

The Complete Beginner's Guide to The World Wide Web

Scott Western, an acknowledged, British, World Wide Web expert, leads you through every aspect of THE WEB highlighting interesting sites, and showing you the best ways to find and retrieve the information you want. Discover:

- ✔ How to minimise your time on-line, saving money for you or your company.
- ✔ Professional tricks for searching the Web
- ✔ How World Wide Web pages are designed and constructed
- ✔ All about domain names and getting your own web space

UK£5.95
US$9.95

Tax Self Assessment Made Easy

Like it or not, the biggest change to the UK tax system has taken place. Self assessment is already in place for many taxpayers who may not even know it. Can it be ignored? No! New requirements for keeping records for example, or changes in the date for submitting tax returns will affect NINE MILLION people according to The Revenue. Penalties for not keeping records can be £3,000, whilst late tax returns can be charged at up to £60 per day.

Thankfully, Stefan Bernstein has distilled all the jargon down to a simple easy to follow guide **at a price the ordinary taxpayer can afford.** The book tells you what you have to do and when to do it, warning you of what happens if you don't.

The Complete Beginner's Guide to Windows 98

An easy-to-read guide to Windows 98 with simple instructions and hundreds of useful illustrations. It leads you through everything from installing Windows 98 to exploring the many exciting features on offer such as the dynamic Active Desktop and revamped Explorer.

Also find out how to access the Internet using the Windows 98 Web browser, email program, newsgroup reader, Web page editor and even a Web publishing wizard. Then learn how to use the amazing "Internet Telephone" that lets you talk over the Net instead of making expensive STD or international calls.

UK£5.95/US$9.95

The Complete Beginner's Guide to Windows 98 uses plain English to explain all Windows 98 has to offer; making it perfect for novices and experienced computer users alike:

- ✪ Master the new and easy-to-use Address Book.
- ✪ Handle multimedia, both live on the Internet and from CD-ROM and DVD. You can even turn your PC into a TV set!

The Complete Beginner's Guide to The Internet

What exactly is The Internet? Where did it come from and where is it going? And, more importantly, how can everybody take their place in this new community?

The Complete Beginner's Guide to The Internet answers all of those questions and more. On top of being an indispensable guide to the basics of Cyberspace,

❑ It is the lowest priced introduction on the market by a long way at a surfer-friendly £4.95. Who wants to spend £30+ on an alternative to find out The Internet is not for them?

❑ It comes in an easy-to-read format. Alternatives, with their 300+ pages, are intimidating even to those who are familiar with The Net, let alone complete beginners!

UK£5.95/US$9.95

The Complete Beginner's Guide to The Internet tells you:

- What types of resources are available for private, educational and business use,
- What software and hardware you need to access them,
- How to communicate with others, and
- The rules of the Superhighway, or 'netiquette'.

Book Order Form

Please complete the form USING BLOCK CAPITALS and return to **TTL, PO Box 200, Harrogate HG1 2YR, England** or fax to **+44 - (0)1423-526035**

❑ I enclose a cheque/postal order for ______________ made payable to '**TTL**'

❑ Please debit my Visa/ Amex/Mastercard No:

Book	Qty	Price
Postage: Free in the U.K., please add £1 ($2) per book elsewhere.		
Total:		

Expiry date:

Signature:

Date:

Title: ______________ Initials: ______________

Name: ______________

Address: ______________

______________ Postcode/Zip: ______________

Daytime Telephone: ______________

Please allow 14-21 days delivery.

We hope to make you further exciting offers in the future. If you do not wish to receive these, please write to us at the above address.

eclipse

When you receive your glasses, and before you use them, please check them VERY carefully for any signs of damage. If they are damaged in any way, do not use them.

To receive your free glasses, simply send the *coupon on the corner of this page* as proof of purchase, along with an S.A.E. to:

Free Solar Eclipse Viewer

Take That Ltd.,

PO Box 200,

Harrogate

HG1 2YR

England

Further pairs of glasses may be obtained at the discounted price of only £1.50 (normally £2) including VAT and postage by sending a cheque/postal order (made payable to TTL) or credit card details to the same address.

* The publisher, author, and their respective employees or agents, shall not accept responsibility for injury, loss or damage occasioned by any person using the solar eclipse viewer whether or not such injury, loss or damage is in any way due to any negligent act or omission, breach of duty or default on the part of the publisher, author, or their respective employees or agents. Glasses are supplied by Peter Allen Eyeware, london.

Free Solar
Eclipse Viewer
Proof of
Purchase

FREE aluminised mylar eclipse viewing glasses

As a purchaser of ***Eclipse: An introduction to total and partial eclipses of the sun and moon*** you may receive one free pair of eclipse viewing glasses totally FREE!

These viewers are made from layers of aluminised mylar bonded together with the aluminium coating on the inside of the layers. This combination of materials reduces the intensity of light emitted from the Sun to a level that is acceptable to the human eye. Using these glasses, you can safely observe a solar eclipse before and after totality.

These viewers have passed stringent testing under EU regulations and carry the CE mark*.

Under NO circumstances should you use these glasses with any other optical devices such as a telescope, binoculars or a camera. They are designed for hand use only, and the intesification produced by optical devices will be too strong for the glasses.

Readers who suffer from any form of eye impediment, disease or who have had eye surgery should get specialist medical advice from their doctor before using the glasses.

⇦ *Continued overleaf*